DESIGNING PLEASURABLE
PRODUCTS

DATE DUE FOR RETURN

Designing
Pleasurable Products
An introduction to
the new human factors

Patrick W. Jordan

Taylor & Francis
Taylor & Francis Group

Boca Raton   London   New York   Singapore

A CRC title, part of the Taylor & Francis imprint, a member of the
Taylor & Francis Group, the academic division of T&F Informa plc.

First published 2000
by Taylor & Francis
11 New Fetter Lane, London EC4P 4EE

Simultaneously published in the USA and Canada
by Taylor & Francis Inc,
325 Chestnut Street, 8th Floor, Philadelphia PA 19106

First published in paperback 2002

Taylor & Francis is an imprint of the Taylor & Francis Group

© 2000 Patrick W. Jordan

Typeset in Times by Taylor & Francis Books Ltd
Printed and bound in Great Britain by
St Edmundsbury Press Ltd, Bury St Edmunds, Suffolk

Every effort has been made to ensure that the advice and
information in this book is true and accurate at the time of
going to press. However, neither the publisher nor the author
can accept any legal responsibility or liability for any errors
or omissions that may be made. In the case of drug
administration, any medical procedure or the use of technical
equipment mentioned within this book, you are strongly
advised to consult the manufacturer's guidelines.

British Library Cataloguing in Publication Data
A catalogue record for this book is available from the British Library

Library of Congress Cataloging in Publication Data
A catalogue record for this book has been requested

ISBN 0–748–40844–4 (Hbk)
ISBN 0–415–29887–3 (Pbk)

CONTENTS

ILLUSTRATIONS

Figures

Tables

ACKNOWLEDGEMENTS

In addition to the literature cited in this book, much of the text draws on my own personal experiences of working in design and human factors. I am grateful to my colleagues within Philips Design who have helped me in the development of many of the ideas contained in the book – I have learned a lot from them since joining the studio six and a half years ago. The numerous awards and publications that we have accrued in that period are testimony to the imagination, inspiration and professionalism of the colleagues that I have had the privilege to work with.

Similarly, I owe a debt of gratitude to many fellow professionals from other design studios, universities and research institutions with whom I have had contact, through conferences or just informally. Again, this has been a valuable learning experience.

Much of the photography in this book is the work of Jim Cockerille (Figures 2.8, 3.5, 3.15, 3.16, 4.7). Jim also provided the creative direction for the cover art. Other photographers who supplied material for the book are Jesper Sandholt (Figures 2.6, 3.4) and Alastair Macdonald (Figures 2.5, 4.8). The photograph in Figure 4.6 was provided courtesy of Good Grips.

Personal thanks to those who helped and encouraged me during the process of writing this book including all at Taylor & Francis.

1

PLEASURE WITH PRODUCTS
Beyond usability

The rise of human factors: usability as a competitive issue

Human factors have come to increased prominence in recent years. This is manifest in a number of ways: one is the ever expanding literature relating to human-factors issues, including books and journals, and even magazine and newspaper articles; another is the number of international conferences and seminars dedicated to human-factors issues. Examples of the latter include the Ergonomics Society Conference in the UK and the Human Factors and Ergonomics Society Conference in the USA. However, perhaps the most important reflection on how seriously human-factors issues are now being taken is the sharp increase in human-factors professionals employed in industry. In particular, human factors are being taken increasingly seriously as an issue within product design. Industrial design departments within most major companies and design consultancies now employ a number of specialists charged with ensuring that product designs fit the needs of those who will use the products. In addition, professionals such as industrial designers and software designers are increasingly expected to have an awareness of human-factors issues and to put them at the centre of the design process.

This has not always been the case. Indeed, the level of integration of human factors within design seems to have gone through three distinct phases, as follows.

Phase 1 – being ignored

Going back fifteen to twenty years, few manufacturing organisations employed human-factors specialists, even amongst the larger companies, and those that did were likely to be involved in defence work. Certainly, human factors were not much of a consideration for companies making consumer products.

Phase 2 – 'bolt-on' human factors

This was the era of creating a new product and then asking the human-factors specialist to help add on a nice interface. The problem with this, of course, was that by this stage in the product-creation process the basic inter-action structure had often been decided, leaving room only for comparatively superficial interface improvements. Nevertheless, this marked an era when more human-factors specialists were finding employment in industry and, although sometimes misunderstood, human-factors issues started attracting attention.

Phase 3 – integrated human factors

And so to the present day. In a number of companies human factors have become seen as something that is inseparable from the design process. Within most major manufacturing companies, product development protocols make provision for the consideration of human-factors issues throughout the design process. This gives the human-factors specialist the chance to influence the design right from its conception.

So why have these changes come about? Probably the main reason is the perceived commercial advantage that good human factors – indeed good design generally – can bring to a manufacturing organisation. In many product areas, technical advances and manufacturing processes have reached a level of sophistication that makes any potential competitive advantage, in terms of functionality, reliability and manufacturing costs, marginal. Many manufacturers now see design as one of the few areas in which it is still possible to gain significant advantages over the competition. Good human factors are, of course, central to achieving excellence in design.

Indeed, customers are becoming increasingly sophisticated in terms of their knowledge of human factors and the quality level of human factors that they expect with a product. Whilst once good human factors may have been seen as a bonus, they are now becoming an expectation. Users are no longer willing to accept difficulties in interacting with products as a price they must pay for 'technical wizardry'. Customers now demand technical wizardry and good human factors and will be antagonised by products that fail to support an adequate quality of use. In the end, of course, such products will also antagonise those who manufacture them as they will find that their customers soon start to look elsewhere (Green and Jordan 1999).

Perhaps the product that brought about the biggest 'sea change' with respect to attitudes towards human factors was the Apple Macintosh computer. Apple first came to prominence as the user-friendly computer company by producing interfaces that relied on direct manipulation, rather than command lines, for executing particular tasks. The interface was, and indeed still is, based on an office metaphor, with icons representing, for example, desks, folders, files and waste-paper bins. The approach, then, was to take an environment that the user was familiar with – the office – and to design the interaction structure within this paradigm. The interface to the Apple Macintosh computer is illustrated in Figure 1.1.

At the time this was a revolutionary step. Previously, software packages had largely relied on command line interfaces. Interacting with these required the learning and memorising of strings of characters and numbers – a task that became more and more demanding as the range of functionality associated with such packages increased. Furthermore, the interaction protocols tended to be 'unforgiving' of error. If the user were to omit even a single character or put a character out of place, the command would not be executed. Rather, the computer would respond with some unhelpful and technical-sounding message – usually 'syntax error'. These old-style interfaces, then, put the technology at the centre of the design process rather than the person using the machine. The user was expected to learn a language whose structure mirrored that of the language used by the professionals who developed and programmed the software. The Macintosh's office metaphor turned things around. Now the computer was speaking the user's language.

This had a profound effect on the way that computer use spread within society – by the end of 1998 there were forty-five computers per hundred Americans and around thirty per European (*The Economist* 1998: 46). Ten to fifteen years earlier virtually nobody owned one. Whilst previously computer operation had been a specialist activity that required extensive learning, now personal computers started to become accessible to the non-specialist – surely a major factor behind the huge growth in computer use in the workplace and at home. It also led to a change in users' attitudes and expectations with respect to personal computing. No longer was computing seen as an opaque and complex activity. Users have come to expect that software packages and operating environments will be both supportive of their needs and simple to use. The development of the Windows environment is further evidence of the software industry's acknowledgement of this change in attitude. It is no longer seen as commercially viable to mass market software that is not designed around a user-centred paradigm.

Changing times: usability as a dissatisfier

Human factors, then, have been seen to add value to products by helping to make them easy to use. However, because customers have come to expect products to be easy to use, usability has moved from being what marketing professionals call a 'satisfier' to being a 'dissatisfier'. In other words, people are no longer pleasantly surprised when a product is usable, but are unpleasantly surprised by difficulty in use. In parallel with this change in attitude

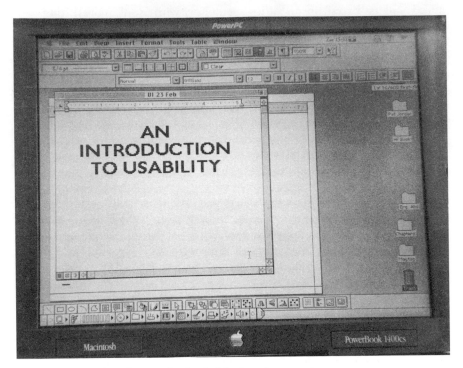

Figure 1.1   The interface to the Apple Macintosh computer

comes a change in the status of human factors within the design process. If the contribution of human factors is simply to enhance usability, then it will come to be seen as a problem-solving discipline, rather than as a discipline that is positively increasing the market value of the products to which it contributes.

That addressing usability issues is a vital role for human factors is not in dispute. However, usability-based approaches are limited. By looking at the relationship between people and products in a more holistic manner, the discipline can contribute far more. Such holistic approaches are known as 'pleasure-based' approaches and are increasingly being adopted by industry-based human-factors professionals – many examples will be given throughout this book. This has led to approaches that look both at and beyond usability. Such approaches have been termed the 'New Human Factors' (Fulton 1993).

Hierarchy of consumer needs

The psychologist Abraham Maslow described a 'hierarchy of human needs' (Maslow 1970). This model viewed the human as a 'wanting animal' who

4

rarely reaches a state of complete satisfaction. Indeed, if a nirvana is reached it will usually only be temporary because once one desire has been fulfilled another will soon surface to take its place. Maslow's hierarchy is illustrated in Figure 1.2. The idea is that as soon as people have fulfilled the needs lower down the hierarchy, they will then want to fulfil the needs higher up. This means that even if basic needs – such as physiological and safety ones – have been met, people will still meet with frustration if their higher goals are not met.

The merits or otherwise of Maslow's theories are not a matter for discussion in this text (a good overview of Maslow's work can be found in Hjelle and Ziegler 1981). The point to note is simply that when people get used to having something, they then start looking for something more. This lesson may apply to human factors as much as to anything else. Taking the idea of a hierarchy of needs and applying it to human factors, the model illustrated in Figure 1.3 is proposed. It is intended to reflect the way that the contribution of human factors to product design might be seen – either explicitly or implicitly – by both manufacturers and those who buy and use their products.

Level 1 – functionality

Clearly, a product will be useless if it does not contain appropriate functionality: a product cannot be usable if it does not contain the functions necessary to perform the tasks for which it is intended. If a product does not have the right functionality this will cause dissatisfaction. In order to be able to fulfil people's needs on this level, those involved in product creation – including, and indeed especially, the human-factors specialist – must have an understanding of what the product will be used for and the context and environment in which it will be used.

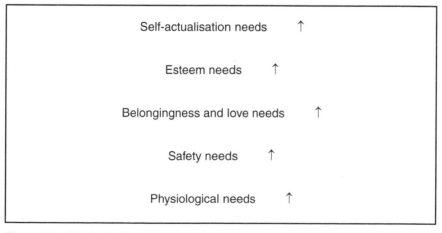

Figure 1.2   Maslow's hierarchy of needs

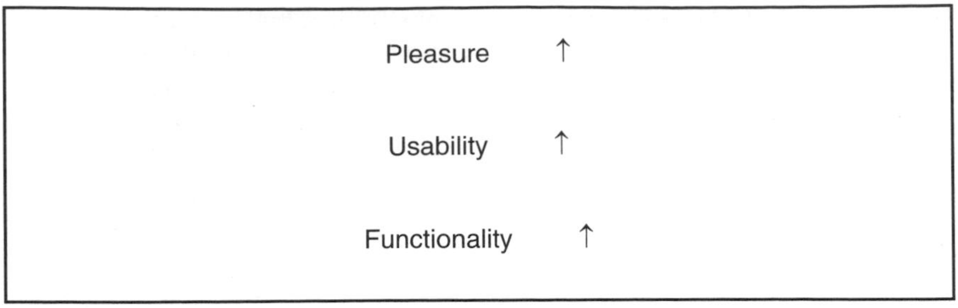

```
Pleasure        ↑

Usability        ↑

Functionality        ↑
```

Figure 1.3   A hierarchy of consumer needs

Level 2 – usability

Once people had become used to having appropriate functionality, they then wanted products that were easy to use. As discussed, this more or less represents the situation at the moment in many product areas: people are used to products that function well, but now they also expect usability. Having appropriate functionality is a prerequisite of usability, but it does not guarantee usability. The human-factors profession is now adept at contributing to the creation of usable products and has established a number of principles to which designs must adhere in order to be easy to use. The basic principles of designing usable products are outlined in *An Introduction to Usability* (Jordan 1998).

Level 3 – pleasure

Having become used to usable products, it seems inevitable that people will soon want something more: products that offer something extra; products that are not merely tools but 'living objects' that people can relate to; products that bring not only functional benefits but also emotional ones. This is the new challenge for human factors.

Usability: vital but not the whole story

It is important to note that pleasure-based approaches to human factors are not an alternative to usability-based approaches. Although usable products will not necessarily be pleasurable, products that are not usable are unlikely to be pleasurable. Usability, then, should be seen, in many cases, as a key component of pleasurability. After all, what is the point of providing a user with a beautiful product with a vast array of functions if the design of the product makes it difficult to use to its full advantage? Nevertheless, usability-based approaches are inherently limited.

The reason why they are limited is that usability-based approaches tend to look at products as *tools* with which *users* complete *tasks*. However, products are not merely tools: they can be seen as *living objects* with which *people* have *relationships*. Products are objects that can make people happy or angry, proud or ashamed, secure or anxious. Products can empower, infuriate or delight – they have personality (Marzano 1998).

People also have personalities. Not only do they have personalities, but they also have hopes, fears, dreams and aspirations. These are liable to affect the way that people respond to and interact with products. Again, this may, prima facie, seem obvious. However, if a being from another planet were to try to learn about the human race via the human-factors literature he/she/it would probably conclude that we were basically little more than cognitive and physical processors. It is a rarity to find published human-factors studies that describe people in terms that go beyond factors such as age, gender, education or profession. Similarly, these studies only seem to be concerned with the level of effectiveness, efficiency and satisfaction with which people can perform tasks – not with their emotional responses to the products that they are using and experiencing. This approach is reflected in the International Standards Organisation's definition of usability: 'the effectiveness, efficiency and satisfaction with which specified users can achieve specified goals in particular environments' (ISO DIS 9241-11).

It might seem that this line of argument is merely a semantic quibble. Surely, it might be argued, the satisfaction component of usability could cover the wider aspects of people–product relationships.

A review of the human-factors literature will make it clear that this is not a mere semantic quibble. The human-factors profession has traditionally operationalised 'satisfaction' in a manner that is limited to the avoidance of physical or cognitive discomfort. This is clearly reflected in the International Standards Organisation's definition of satisfaction: 'the level of comfort that the user feels when using a product and how acceptable the product is as a vehicle for achieving their goals' (ISO DIS 9241-11).

This, then, is the problem with usability-based approaches – they tend to encourage the view that users are merely cognitive and physical components of a system consisting of the user, the product and the environment of use. The premise on which these approaches are based appears to be that the product must be designed such that the cognitive and physical demands placed on the users are minimised – that the demands do not exceed the person's processing capacity. Usability-based approaches then encourage a limited view of the person using the product. This is – by implication if not by intention – dehumanising.

This seems ironic; of all the people involved in the product- creation process, it is the human-factors specialist, with his or her roots in behavioural science, who is the person that would be expected to have the richest understanding of the people for whom the products are being created. It is no longer sufficient for the profession to think of people in such limited terms. In order to fully represent, in the product-creation process, the people using and experiencing the product, human-factors specialists must take a wider view of person-centred design and look, in a more holistic context, both at product use and at those using and experiencing products.

Pleasure-based approaches: challenges for the new human factors

So far in this chapter it has been argued that human factors must move beyond usability in order to address the relationship between people and products holistically. This presents the profession with a number of new challenges that must be addressed in order to support such approaches. Perhaps the three major challenges are as follows.

Understanding people holistically

Meeting this challenge requires going beyond simply looking at the factors that influence how successful – in terms of task completion – a user–product interaction will be. In order to find a way into the wider issues of people–product relationships, it is necessary not only to have an understanding of how people use products, but also of the wider role that products play in people's lives. Such an understanding is a precondition of being able to define a product benefits specification, which goes beyond the traditional usability-based user requirements specification.

Linking product benefits to product properties

Having established the different types of benefits – or 'pleasures' – that people can derive from products, the next stage is to link those pleasures to particular aspects of a product design. For example, it might be that feelings of security with a product are linked to high levels of usability and/or high levels of product reliability. A feeling of pride may be linked to, say, good aesthetics. Similarly, particular types of displeasure may also be linked to inadequacies with respect to certain product properties. For example, annoyance might be linked to poor technical performance whilst anxiety may be related to a lack of usability.

All of the above suggestions of links between pleasures and product properties are speculations. This is an issue that the human factors must address systematically if it is to make a significant contribution to the design and development of pleasurable products.

Developing methods and metrics for assessing product pleasurability

Creating pleasurable products, like most good commercial design, will be dependent on the iterative development of design concepts. This means creating prototypes of designs and evaluating them to see if they promise to deliver the pleasures intended. Being able to do this relies on having the necessary methods and metrics available in order to assess the pleasurability of designs and of design prototypes.

The aim and contents of the book

The aim of this book is to provide the reader with an introduction to pleasure-based approaches to human factors in product design. In Chapter 2 the reader will be introduced to the concept of pleasure with products. A definition of pleasure with products will be offered and a practical framework for addressing issues of product pleasurability will be outlined.

Chapter 3 will address the three challenges outlined above: understanding people, linking product benefits to product properties and assessing product pleasurability. Essentially, the chapter will outline a process for creating pleasurable products. This chapter will be complemented by Chapter 4, which describes a number of methods that can be used in the product-creation process. Case studies will be reported, demonstrating how these methods have been used in designing pleasurable products. Meanwhile, in the last chapter – Chapter 5 – the reader will be briefly reminded of some of the major issues covered by the book and a few conclusions will be drawn.

Throughout the book, the issues addressed will be illustrated with examples and case studies. These will be taken from across the whole spectrum of design, and the application areas covered will include: consumer products; professional products; computer software and hardware; graphic design; vehicle design; fashion design; furniture design; packaging; and architecture.

The scope of the book and its limitations

It is hoped, first, that by the end of the book readers will have gained an overview of the role of human factors in product design; and, second, that they will understand what must be taken into account in order to design products that will be pleasurable for those who own and use them. No previous knowledge of human-factors issues is assumed and the book should be accessible to all, regardless of academic or professional background.

This book is not a 'manual'. Reading the book will not give a person new skills – for example it will not turn a marketing professional into someone who can design pleasurable products or a software designer into a specialist at evaluation. Nevertheless, it should give all readers an awareness of the issues associated with creating pleasurable products and will give those who already have particular skills the knowledge to enhance their effectiveness. For example, those who already have experience of usability-based approaches to product creation should be able to adapt and extend these approaches in order to contribute to the creation of more broadly pleasurable products.

Who is the book for?

This book is written primarily for students and professionals involved in product design. 'Product', as used in this text, is a generic term covering, for example, consumer products, professional products, computer software, motor vehicles and manufacturing equipment and machinery. Indeed, the principles and approaches outlined should be applicable over a wide array of application areas. Courses for which this book should be suitable include: human factors/ergonomics; industrial design; human–computer interaction; graphic design; information technology; market research; marketing; multimedia; and modules on human–computer interaction in psychology, computer science and engineering courses. Professions at which the book is aimed include: human factors; interaction design; industrial design; software design; marketing; market research; product management; and engineering.

2

THE FOUR PLEASURES

In the previous chapter it was argued that human factors should move beyond usability-based approaches to design. It was suggested that such approaches are limited, even dehumanising, as they tend to focus merely on the fit of a product to a person's cognitive and physical characteristics. Pleasure-based approaches, on the other hand, encourage a holistic view of the user, judging the quality of a design on the basis of the wider relationship between a product and the people for whom it is designed.

In this chapter the various aspects of people–product relationships are explored. A framework is given – known as the 'four pleasures' – within which to consider the different types of pleasure that people may seek and that products may bring. The application of this framework to understanding people is illustrated with an example that demonstrates the difference between pleasure-based and usability-based approaches to defining the characteristics of people. The framework is then used as a means of structuring a set of examples, demonstrating a variety of ways in which products can give pleasure to those who experience them.

In search of pleasure

Since the beginning of time humans have sought pleasure. We have gained pleasure from the natural environment: from the beauty of flowers or the feeling of the sun on our skin; from bathing in soothing waters or the refreshment of a cool breeze. We have actively sought pleasure, creating activities and pastimes to stretch our mental and physical capabilities or to express our creative capabilities. Cave dwellers wrestled to test their strength and expressed themselves through painting on the walls of their dwellings. Today we 'pump iron' in the gymnasium and decorate our homes with selections of paintings and posters.

Another source of pleasure has been the artefacts with which we have surrounded ourselves. For centuries humans have sought to create functional and decorative artefacts – artefacts that have increased the quality of life and brought pleasure to the owners and users. Originally, these objects would have been clumsily bashed out from stone, bronze or iron by unskilled people who simply wanted to make something for their own use. As systems of trade and barter were developed, specialist craftspeople became prevalent, creating artefacts for use by others in the community. Today, most of the artefacts that we surround ourselves with were created by industry.

Pleasure

What is pleasure? The *Oxford English Dictionary* defines it as 'the condition of consciousness or sensation induced by the enjoyment or anticipation of what is felt or viewed as good or desirable; enjoyment, delight, gratification. The opposite of pain.' In the context of products, pleasure can be defined as 'Pleasure with products: The emotional, hedonic and practical benefits associated with products' (Jordan 1999).

Practical benefits are those that accrue from the outcomes of tasks for which the product is used. For example, in the context of a word-processing package, a practical benefit could be the effective and efficient production of neat, well-presented documents. Meanwhile, a washing machine, for example, delivers the practical benefit of clean, fresh clothes.

Emotional benefits are those pertaining to how a product affects a person's mood. Using a product might be, for example, exciting, interesting, fun, satisfying or confidence enhancing. A computer game, for example, might be exciting and fun to use, whilst a stylish new dress may give the wearer a feeling of self-confidence.

Hedonic benefits are those pertaining to the sensory and aesthetic pleasures associated with products. For example, a person might recognise a product as an object of beauty or may enjoy the physical sensation of touching or holding a particular product. A well-designed chair, for example, may be physically comfortable to sit on and may also be an *objet d'art* worthy of aesthetic appreciation. Meanwhile, a shaver might give pleasant tactile feedback, both in the hand and on the face.

In a sense this definition is a 'catch all'. Indeed, it is intended as such. Pleasure-based approaches to product design can be seen as approaches that consider the *all* of the potential benefits that a product can deliver.

It is important to note that pleasure with products accrues from the relationship between a person and a product. Pleasurability, then, is not simply a property of a product but of the interaction between a product and a person. Whilst one person might derive pleasure from, for example, the status associated with carrying around a personal organiser, another person might consider ownership of such a product to be pretentious and thus embarrassing. The characteristics of people that affect how they relate to products will be discussed in the context of many of the examples discussed later in this chapter. They will also be considered, in a more systematic manner, in Chapter 3.

The four pleasures: a framework for considering pleasure with products

A useful way of classifying different types of pleasure has been espoused by Canadian anthropologist Lionel Tiger. Tiger has made an extensive study of pleasure and has developed a framework for addressing pleasure issues, which he outlines in some depth in the book *The Pursuit of Pleasure* (Tiger 1992). The framework models four conceptually distinct types of pleasure – physical, social, psychological and ideological. Summaries of Tiger's descriptions of each are given below. Examples are added to demonstrate how each of these components might be relevant in the context of products. This framework will be used throughout this book as a means of structuring thought with respect to pleasure with products.

Physio-pleasure

This is to do with the body and with pleasures derived from the sensory organs. They include pleasures connected with touch, taste and smell as well as feelings of sensual pleasure. In the context of products, physio-pleasure would cover, for example, tactile and olfactory properties. Tactile pleasures concern holding and touching a product during interaction. This might be relevant, for example, in the context of a telephone handset or a remote control. Olfactory pleasures concern the smell of the new product. For example, the smell inside a new car may be a factor that affects how pleasurable it is for the owner.

Socio-pleasure

This is the enjoyment derived from relationships with others. This might mean relationships with friends and loved ones, with colleagues or with like-minded people. However, it might also include a person's relationship with society as a whole – issues such as status and image may play a role here.

Products can facilitate social interaction in a number of ways. For example, a coffee-maker provides a service that can act as a focal point for a little social gathering – a 'coffee morning'. Part of the pleasure of hosting a coffee morning may come from the efficient provision of well-made coffee to the guests. Other products may facilitate social interaction by being talking points in themselves. For example, a special piece of jewellery may attract comment, as may an interesting household product, such as an unusually styled television set. Association with other types of products may indicate belonging to a social group: Porsches for 'yuppies'; Dr Marten's boots for skinheads. Here, the person's relationship with the product forms part of their social identity.

13

Psycho-pleasure

Psycho-pleasure pertains to people's cognitive and emotional reactions. In the case of products, this might include issues relating to the cognitive demands of using the product and the emotional reactions engendered through experiencing the product. For example, it might be expected that a word processor that facilitated quick and easy accomplishment of, say, formatting tasks would provide a higher level of psycho-pleasure than one with which the user was likely to make many errors. The former word processor should enable the user to complete the task more easily than he or she would with the latter. The outcome may also be more emotionally satisfying.

Ideo-pleasure

Ideo-pleasure pertains to people's values. Tiger (1992) refers to the pleasures derived from 'theoretical' entities such as books, music and art. In the context of products it would relate to, for example, the aesthetics of a product and the values that a product embodies. For example, a product made from bio-degradable materials might be seen as embodying the value of environmental responsibility. This, then, would be a potential source of ideo-pleasure to those who are particularly concerned about environmental issues. Ideo-pleasure would also include the idea of products as art forms. For example, the video cassette recorder that someone has in the home is not only a functional item, but something that the owner and others will see every time that they enter the room. The level of pleasure given by the video cassette recorder may, then, be highly dependent on how it affects its environment aesthetically.

Need pleasures and pleasures of appreciation

In his treatise *The Four Loves* (Lewis 1960), the philosopher and scholar C.S. Lewis includes an essay on what he describes as likings and loves for the sub-human. In this essay he considers natural entities, such as plants and animals, but also artefacts, such as products. Lewis classifies the pleasures that can be derived from such entities as being either 'need pleasures' or 'pleasures of appreciation'.

Broadly, need pleasures can be seen as pleasures that accrue by moving a person from a state of discontentment to one of contentment: for example, drinking a glass of water would give a need pleasure to someone who was thirsty. Pleasures of appreciation, meanwhile, are those that accrue because a person finds something positively pleasurable, no matter what their current level of contentment. In these cases the person feels that the entity being encountered is worthy of appreciation either in its own right or because of some additional pleasure that it delivers. On one hand, a person might enjoy a fine wine for its taste and bouquet and for the pleasant feeling of intoxication that it delivers, no matter if he or she initially felt in need of intoxication or of quenching a thirst. On the other hand, it seems unlikely that someone would wish to drink a glass of water unless they were already thirsty.

The important thing to note, then, is that pleasure can be thought of both as the elimination of, or absence of, pain and also as the provision of positive, joyful feelings. Those involved in the creation of pleasurable products should be aware of both of these aspects as issues to be addressed in the product creation process.

A note on the four pleasures

It is important to note that the use of the four-pleasure framework is simply intended as a means of structuring thought as regards pleasure (Tiger 1992). It is not intended, in itself, to give an insight into *why* people experience pleasure. The benefit that use of the framework provides comes through breaking the issue of pleasure with products into four sections. Doing so can help to make it easier for those involved in the design process to consider the full spectrum of the sorts of pleasures that a product can bring. The four-pleasure framework is not, then, a theory of pleasure, but simply a tool that can help in taking a structured approach to the issue.

Whilst it may be useful to consider all four types of pleasure when approaching the issue of how a product can please those for whom it is designed, there is no suggestion that all products should provide all four types of pleasure. It might be that the benefits associated with a particular product cover the entire range of different types of pleasure, or that a product is experienced as pleasurable in one particular way. This should become clear from the examples given later in this chapter. Similarly, there will also be pleasures that are difficult to classify in terms of which one of the four pleasure categories they fall under. Again, this need not be a problem. Designing a product so that it delivers a particular benefit to the people experiencing it does not depend on knowing which category of pleasure that benefit falls under. Rather, what is important is that the benefit is identified in the first place. The structured approach facilitated through use of the framework can help to ensure that possible benefits are not overlooked.

Understanding people holistically

Most of the rest of this chapter is given over to examples illustrating the different types of pleasure that products can give. These are considered in the context of the four-pleasure framework. Before that, in this section, an example is given showing how the four-pleasure framework can be used to help understand people holistically and to demonstrate that pleasure-based approaches to human factors rely on building up a far richer understanding of people than do usability-based approaches.

Background: Janet Peters

The imaginary subject of this example is Janet Peters, a 23-year-old accountant living in Reading, near London. Janet has been working within a major accountancy consultancy for the past year; her first job since leaving university, where she obtained a second-class honours degree in accountancy. She is very ambitious and wants to get to the top as quickly as possible. She is quite prepared to put in long hours at the office, because she feels that hard work is the key to success.

Nevertheless, work is not her whole life: she is a keen sports player – an active member of the local netball team. She not only enjoys netball matches for the fun of the game, but also because she knows that it helps to keep her fit, providing a good balance to the sedentary nature of her job. The camaraderie of the team also appeals to her, and in particular she enjoys the high spirits of the after-match drinking session!

Janet is dating Mark, a 25-year-old stockbroker from London. She is very fond of Mark and they have a lot in common – they are both ambitious people and love the trappings of a yuppie lifestyle. Janet spends most weekends at Mark's flat in Kensington, an upmarket district of London. They love eating out at exclusive restaurants and going to shows in the West End.

Janet keeps in regular contact with her parents, phoning them at least twice a week. She loves to tell them all about her work and the wonderful lifestyle she leads. They are very proud of what she has achieved – they had always wanted her to have the opportunities in life that they never had. Her parents live in a quiet country village in Dorset, a rural area in the south-west of England. Janet still has many friends there, and she sees them a couple of times a year when she goes home to visit her parents. Her friends are slightly in awe of her glamorous lifestyle. Janet knows this and takes a secret pleasure in it – she is proud of what she has so far achieved and knows that she can be even more successful in the future.

So, that is some basic background to Janet's life. In the next sections, an analysis of Janet is outlined, based on the structure provided by the four-pleasure framework.

Physio-pleasures

From what is known about Janet, it is clear that she has a job that is physically undemanding and that she balances this with a leisure activity – netball – that is physically demanding. So, what are the pleasures of need and appreciation that might be associated with this?

One of the reasons that she plays netball is because it helps her to stay in shape. For Janet, staying in shape is a need pleasure. If she is out of shape, she feels miserable and unattractive – staying in shape is, for her, a prerequisite of feeling okay. Clearly, being in or out of shape is not something that changes on a day-to-day basis, but is the result of on-going lifestyle activities. Nevertheless, she knows that every little bit makes a difference: every hour she spends sitting at her desk is a negative influence and every game of netball is correspondingly positive.

Sometimes, Janet likes to spoil herself with luxuriant physical relaxation. If she is spending the evening alone she loves to soak in a long hot bath or to curl up on the sofa with a bottle of wine and watch some undemanding drama on television. For Janet these are pleasures of appreciation – she would still be content even without the feel of the warm water on her skin or the intoxicating effect of the wine. These, then, are, for her, some of the extra-special things that make life wonderful.

Socio-pleasures

Janet loves the glamour of her lifestyle and takes a secret pleasure in the status associated with it. She loves being the envy of her friends and knows how proud her parents are of her. These pleasures are, for Janet, ones of appreciation. She would still be content without such glamour – the status and her parents' pride are positive benefits to be enjoyed.

Janet's basic social needs are to be loved and to feel a sense of belonging, otherwise she can feel very lonely. Because of this her relationships with her parents and with Mark are very important. It is also very important for her to make new friends so that she can establish a social life for herself in Reading. She is concerned that, because she spends most of her weekends in London, she is not building up much of a life for herself in Reading. These, then, are Janet's social need pleasures – if she does not have good relationships with her boyfriend and her parents she will feel discontented.

Psycho-pleasures

Because her job can be very demanding, Janet can be prone to stress. This, of course, makes her unhappy, so the relief of stress, for Janet, is a psycho-need pleasure. When she is not working, Janet is the sort of person who is easily bored. Another need pleasure, for her, is the need to partake in hobbies and activities that keep her stimulated.

Sometimes, these activities simply supply relief from boredom, whilst at other times they can provide a real sense of achievement. For example, Janet has recently taken to cooking as a hobby and takes a great deal of pride in cooking a wonderful meal for dinner guests. She is also a reasonably good guitar player and sometimes writes her own songs – again, she gets a real sense of achievement if she feels that she has written something particularly good. Gaining a sense of achievement, then, is a pleasure of appreciation for Janet.

Ideo-pleasures

Janet has a reasonably sophisticated sense of style. She is not a design connoisseur, but she has a view on what good taste is and would feel uncomfortable owning anything that did not conform to her ideas of good taste. For example, after she had just moved into her new flat, a friend had given her a lamp to go in the living room. Although she felt obliged to have it in the room, she found it a rather tacky object and always felt slightly uncomfortable whenever she saw it. Perhaps because of her rural upbringing, Janet is always wary of being seen as unsophisticated. The need to be perceived as a woman of good taste is, then, a need pleasure for her.

Aside from her basic need to be seen as having good taste, Janet also aspires to be seen as a successful young woman. She enjoys looking back at what she has achieved in her career so far and finds her success to be very self-affirming. Her long-term aspirations are to achieve even more career success – to get to a senior position within her profession. For Janet, achieving such aspirations would be pleasures of appreciation. Provided her job is fulfilling she will be contented – notable success would be something to be enjoyed on top of this contentment.

In terms of her moral value system, Janet is fairly liberal. She does not hold any religious beliefs; however, she does feel a general responsibility to treat others decently and to take a degree of responsibility for what is going on in the world around her. Although she will not actively seek out opportunities to exert her principles she will be aware of her principles when confronted by a moral choice. For example, although she is not an active campaigner on environmental issues, she will still tend to prefer products that are environmentally friendly and would feel uncomfortable buying something that she felt was environmentally harmful. To this extent, then, taking environmental responsibility is a need pleasure for her – she will feel uncomfortable if she ignores this responsibility.

Whilst she doesn't take much of a leading role with respect to moral issues, Janet has a great deal of admiration for those who do. For example, she hugely admires great political figures, such as Nelson Mandela, and those who dedicate their lives to helping others, such as Mother Teresa. On a more day-to-day level, Janet admires certain ethical businesses, such as the Body Shop. Indeed, supporting such businesses through buying their goods is a pleasure of appreciation for her.

Summary

The above example shows how the four-pleasure framework can be used as a means of structuring the pleasure issues associated with a particular person. They are summarised in Table 2.1.

Comparison with usability-based approaches to understanding people

This example is based on speculations about a fictitious person. Nevertheless, the issues raised here are indicative of the way in which new human-factors approaches seek to understand people. How to gather such information in order to build up an accurate picture of people will be discussed in the following chapter. The point to note for now is how much richer a picture of a person is provided by this sort of analysis, as compared with the profiles typically gleaned from old-style user analyses.

A traditional human-factors analysis of Janet Peters would have noted that she was young, fit and healthy, and that she had no physical or cognitive disabilities. It might have noted that she was female – if the analyst believed that this would have any cognitive or physical implications – but that would have been about it. Little or no attention would have been paid to her lifestyle or her aspirations, nor would her values have been seen as much of an issue. Again, it will be demonstrated in the following chapter how holistic knowledge about a person's lifestyle and values can be used as the basis for creating a Product Benefits Specification, which in turn can be used as the basis for making design decisions. The remainder of this chapter, meanwhile, will be filled with examples demonstrating how products can deliver various different types of pleasure to the user. Again, this will be done within the context of the four-pleasure framework.

Examples of pleasure with products

In the following sections, a number of examples of product benefits will be

Table 2.1 Four-pleasure analysis for 'Janet Peters'

Physio	*Socio*	*Psycho*	*Ideo*
• Staying in shape	• Company	• Stress relief	• Decency
• Physical relaxation	• Good personal relationships	• Stimulation	• Responsibility
• Intoxication (alcohol)	• Glamour	• Sense of achievement	• Supporting moral leaders
	• Status		

given. The aims are to demonstrate the sorts of pleasures associated with each category of the four-pleasure framework and to give an indication of the wide range of pleasures that products can provide.

Physio-pleasures

Traditional human-factors approaches are very often concerned with physical aspects of product use: for example, with whether the dimensions of the product are suited to the anthropometrics of the users; with whether the product is light enough to carry; with whether the product is usable by people with particular physical disabilities. These sorts of issues might be thought of as physiological need pleasures. In other words, they pertain to physiological aspects of product use that, if handled badly in the design, will make the user feel discontented. If they are handled well in the product design, the user may feel relieved at the absence of problems or may simply not notice.

Consider, for example, the layout of a vehicle interior. The driver will, of course, be seated behind the steering wheel, from which position he or she will have to operate a number of different controls. These will include, for example, controls to operate the in-car stereo, controls for the air conditioning, controls to change the speed and gear ratio of the engine and controls for the turn signal. If the driver cannot easily reach all of these controls then he or she will experience difficulties that might have consequences ranging from discomfort (for example, if he or she is unable to adjust the air conditioning) to danger (for example, if he or she is distracted from driving in the course of reaching for an awkwardly placed control).

Other products have to accommodate people's bodily dimensions. These include medical products (such as X-ray machines and scanners), personal care products (such as sun beds) and, perhaps most obviously of all, chairs and seats.

Many people in the Western world spend the majority of their waking hours seated. People sit in their workplaces, sit in their cars, sit at home to eat and watch television, sit in bars, cinemas, sports arenas, theatres, etc. The two basic need pleasures associated with seating are that it be comfortable and that the person sitting should not be physically damaged by the seat.

When investigating issues connected with comfort in seating, 'comfort' tends to be operationalised as an absence of discomfort. Investigators ask seat testers if they are feeling any discomfort in any parts of their body when seated in a particular seat – if they are, then the seat is deemed to be uncomfortable; if not, the seat is deemed to be comfortable. There is much excellent human-factors literature on how to design comfortable seating – see, for example, *Fitting the Task to the Man* (Grandjean 1988), which gives an overview of issues pertaining to workplace seating. The basic design rule for creating comfortable seating seems to be to make seating as adjustable as possible. Aspects such as the height of the seat from the floor, the angle of the back-rest and the height of the arms – all these should be adjustable. People come in all different shapes and sizes, and seats should be adaptable to this.

Adjustability is also the key to avoiding physical injury from using seating (Grandjean 1988). Unfortunately, back problems are rife in the Western world and seating design is often cited as a major cause of this. This can occur either because seating is not sufficiently adjustable to fit the user or because the user doesn't adjust the seating properly. The same may be true of the way that people use car seating. Many people do not adjust their car seats in the optimal manner – either because they do not know how to or because they do not think it is important.

Of course, the design of the seat and the design of the controls for adjusting the seat are likely to affect whether people are aware that the seat is adjustable and whether they can adjust it properly and easily. An excellent example of a well-designed seating control is the BMW car seat control, as described by Norman (1988). This control is designed to look like the car seat itself and to work in such a way that moving the control will similarly adjust the position of the seat. For example, tipping the control backwards will adjust the seat so that it also tips backwards. This is an excellent illustration of the principles of explicitness and compatibility – both key principles of designing for usability. The control is explicit because it makes it clear what its function is and what adjustment options are available. The control is also compatible with expectations of how people would expect it to work. Tipping it back makes the seat tip back – just what would be expected.

Another approach to relieving possible back problems associated with sitting is to create seating that takes the physical burden off the back and transfers the sitter's weight differently across the body. Perhaps the most acclaimed example of this is the Balans variable seat created by Norwegian designer Peter Opsvik. This design spreads the burden of supporting the sitter's weight across the sitter's knees and the base of his or her spine. This contrasts with conventional seating design, which can put an inordinate burden on the base of the spine – a major cause of the associated back problems.

One of the earliest systematic studies of seating was carried out in the context of the design of aircraft interiors. In 1955, Walter Dorwin Teague was commissioned to redesign the interior of the Boeing 707 (Dormer 1993). This plane had initially been designed as a refuelling tanker for the USA Air Force. Pan-Am airways decided that they would adapt it for use as a passenger carrier. As part of his approach to this redesign, Teague created a full-size mock-up of the interior and conducted 'pretend flights'. This involved asking people to sit in the seats for a length of time and then asking them whether they felt any discomfort. Teague also checked that the dimensions of the interior as a whole fitted around the dimensions of the passengers;

for example, he arranged the overhead luggage lockers such that people did not bang their heads against them when they stood up.

Later work on the interior of the 707 was conducted by Henry Dreyfuss. Dreyfuss was probably the first major designer to rely heavily on the use of anthropometric data. Indeed, he collected a great deal of it himself, which he published in a series of charts. Many designers consider Dreyfuss might be considered as the 'father' of ergonomics in product design.

An excellent overview of issues pertaining to designing products to fit people's body dimensions can be found in *Bodyspace* by Stephen Pheasant (1986). This book gives a thorough introduction to anthropometrics and includes detailed tables summarising the physical dimensions of different populations. When considering anthropometrics and product design, the received wisdom is that the product's critical dimensions should be such that, at minimum, users with dimensions ranging from the fifth percentile to the ninety-fifth percentile of the user population should be catered for. This means that, in other words, if anyone at all has to be excluded from comfortable use of the product it should only be those who are extremely small or extremely large with respect to a particular physical dimension.

In the case of (nominally) portable products, the weight of the product will be an issue affecting how pleasurable it is. Strength can vary widely between different people – products that are portable for one person are not necessarily portable for another. This seems a very straightforward issue, yet it also seems to be one that is often overlooked. For example, many televisions that are sold as 'portables' would be far too heavy for many of the elderly to carry – the weight of these products, then, is excluding many of the potential user group from receiving the benefits of portability. Sometimes there may be an element of danger associated with portable products. For example, kettles have to be moved from one place to another when full of boiling water – dropping a kettle full of scalding water could be very dangerous.

Some portable products are moved across surfaces rather than actually lifted and carried by the user. Many vacuum cleaners are examples of this. Here, then, the issue will be not so much how heavy the product is, but how manoeuvrable it is. Most vacuum cleaners are fitted with wheels and are pulled or pushed by the user. It is important that these products can be manoeuvred without being bumped against other household objects as this may damage the product and, probably more importantly, the object against which it is bumped. One way of making vacuum cleaners more manoeuvrable is to put a swivel-wheel at the front. This makes the product much more responsive to the user's control and has been used as a selling feature for a number of vacuum cleaners

Hand-operated products are another potential source of user discomfort. In the 1950s, Finnish designer Olaf Backstrom made a study of scissors design. He noted, for example, that scissors were often designed in such a way that those using them for long periods of time – for example, tailors, designers and artists – could end up with painful calluses and blisters on their hands. In response to this he created the O-Series range of scissors for Fiskars in 1960. These scissors had a long rectangular handle that users gripped from the outside, plus one finger hole that enabled the user to keep his or her hand firmly located in place. This contrasted with traditional scissors design that encouraged the user to stuff a number of fingers into each finger hole, creating pressure points between the user's fingers and the edge of the handle – this can lead to blistering.

Similarly Zdenek Korvar, a Czech sculptor and designer, carried out extensive work on the ergonomics of hand tools at around the same time. He was interested in understanding the causes of cuts and blisters on factory workers' hands. He set up user trials with workers using existing tools (hammers, pneumatic drills, etc.). Korvar wrapped a soft plaster sheath around the outside of the products before giving them to the workers to use. After use he looked at the imprints made by the workers' hands. He used this as the basis for designing grips that fitted the hand snugly when the products were in use.

Korvar's approach of modelling the product grip to the users' hands has been reflected in the design of many of the recent generations of Japanese hand-held video cameras. These have been designed so that they grip the user's hand firmly and snugly, enabling him or her to move the camera around as if it were almost an extension of the body. Dormer (1993) describes these cameras as looking as if they have evolved biologically rather than having been fabricated. He suggests that they look as though they could almost be implanted into the body, rather than simply being held in the hand.

Sometimes a product will provide people with a physiological need pleasure by protecting them in a potentially dangerous situation. Clothes and accessories may be designed to give protection against, for example, falling objects (e.g. hard hats, steel-toed shoes), heat and chemicals (e.g. protective clothing), accidents (e.g. motorcycle leathers and helmets) and attack (e.g. camouflage clothing, chemical warfare suits).

An important issue to consider when designing protective clothing is that the item being worn should not solve one problem only to introduce another. This might be the case, for example, with safety gloves that will protect the wearer against a spillage of a corrosive substance, yet at the same time are so cumbersome to wear that they make it difficult for the wearer to handle that substance, thus increasing the chance that a spillage will occur in the first place. This sort of criticism has been levelled at full-face motorcycle helmets. Whilst these offer better protection than an open-face helmet in the event of an accident, they offer less all-round visibility to the motorcyclist than does an open-faced helmet. Thus, it could be argued that wearing a full-face helmet increases the chance of an accident occurring.

As another example of this phenomenon, consider the example of ear protectors in the workplace. These can, of course, offer protection from ear damage by limiting the amount of noise reaching the ear – useful, for example, in a working environment in which there is a lot of loud machinery. Unfortunately, however, they present a secondary danger to the wearer by also cutting out sounds that might give crucial cues to the presence of danger. For example, the wearer might not hear an alarm signal or a shout from a colleague in the event of danger occurring. He or she might also be unable to hear critical safety cues about what is going on around him or her. For example, if a fork lift truck were about to come around the corner he or she might not hear it in time to get out of the way.

Protective clothing should also be comfortable, or else people are likely to avoid wearing it. Workers may remove ear protectors if they are uncomfortable to wear. Similarly, people may be disinclined to wear protective suits if they leave the wearer hot and uncomfortable.

Indeed, comfort is generally an issue central to the design of clothing, although this has perhaps only recently been the case. Going back to the turn of this century, there seemed to be a fashion for wearing clothes that would have been distinctly uncomfortable, but which were regarded as beauty-enhancing – the corset being the epitome of such a garment. In recent times, however, attitudes towards comfort have become very different. For example, in the late 1970s and early 1980s tracksuits and training shoes had become part of mainstream (or 'high-street') fashion, largely because they were such comfortable leisurewear. This style was later adopted by designers of high (or 'catwalk') fashion. Examples of the leisurewear influence on catwalk fashion can be seen in Michiko Hoshono's Spring/Summer Collection from 1994 and, in the same year, Donna Karan's DKYN Diffusion Line, which incorporated elements of American sportswear.

If comfort is defined – at least operationally – as an absence of discomfort, then it is clear that products can offer physiological pleasures that go beyond comfort and into the realms of sensuality. In other words, products can provide the user with sensations that are physically pleasurable in and of themselves, even if the user felt perfectly comfortable in the first place. Continuing with the example of clothing, there are certain materials that feel particularly pleasant against the skin and which can provide a positive sensual experience to the wearer. One example is silk: a silk shirt can feel very pleasant against the skin.

Hand-held products can also be sensually pleasurable to the touch. This sensual experience is shaped not only by the form of the product but also by the properties of the materials used. Figure 2.1 illustrates a shaver designed by Philips Design for Philishave. This product is pleasurable to hold both because of its pleasing organic shape and because of the use of rubber-like silicon in combination with the matt plastic finish that forms the main part of the body.

In the case of a product such as a shaver, it is also important that the product feels nice against the face – perhaps, in the case of shavers, this might be seen as a need pleasure. After all, soreness, nicks or cuts will leave the user less than contented. Nevertheless, there are some products whose application to the body can be positively pleasurable. For example, a number of manufacturers are now offering massage products. These take the form of vests to be worn or pads that are placed against the body. They include mechanisms that turn – exerting a gentle but firm pressure against the body.

Other products than can be sensually pleasing to the touch include those where the user interacts through push buttons. The rubber-like silicon used for the buttons on many television remote controls, for example, can be pleasurable to the touch, giving as they do a soft but firm feeling against the user's fingertips (see Figure 2.2).

Whether a particular sensual experience will be regarded as pleasurable may be dependent upon the context in which the experience occurs. For example, a feeling that is considered pleasurable with respect to remote-control buttons might not necessarily be pleasurable in the context of, for example, alphanumeric keyboards or professional products. In the 1960s, Italian designer Mario Bellini spent a number of years researching keyboard design for Olivetti in conjunction with Perry King from the UK and Santiago Miranda from Spain. After a number of experiments, Bellini designed the Divisumma electronic calculator in 1972. This calculator was notable for its continuous rubber membrane keyboard. In 1987, the Museum of Modern Art in New York held a retrospective of Bellini's work. In the catalogue that accompanied the exhibition, it was suggested that the product had been designed more with the intention of creating something

Figure 2.1 Philishave electric shaver

Figure 2.2 Remote control with push
buttons made from rubber-
like silicon

that would be sensorially pleasing to the touch, rather than as something
that was primarily a calculating apparatus. In his critique of post-war
design, Dormer (1993) includes a brief discussion of this product. He points
out that many might regard the first duty of such a machine as being to
perform calculations and thus that the design should primarily be supportive
of ease of use, rather than of tactile sensuality.

Nevertheless, Olivetti believed that this was the way forward and encouraged their designers to use a similar approach in the creation of push
buttons for other professional products. However, many people wanted a
more mechanical feedback, preferring a feeling of precision over the inaccurate feel of squidgy buttons. This is borne out, for example, by the pleasing
feel of the keys on many laptop computers. On the Apple Macintosh
PowerBook, for example, the keys are very light to the touch, but because
they give a gentle, yet clearly audible, click when pushed, they give a feeling

of precision and also help the user to feel that he or she is in control of the situation. This keyboard is illustrated in Figure 2.3.

Pens and pencils are another example of products where designers and manufacturers have experimented with materials and finishes aimed at providing the user with a pleasant sensual experience. Again, manufacturers have tried coatings such as rubbers and silicon composites in order to make the user's tactile experience a pleasurable one. An example is illustrated in Figure 2.4. At the base of the pencil a 'grippy' silicon–rubber composite surround gives the pencil appealing sensorial qualities and gives the user a feeling of control.

Sometimes a product will provide a sensual pleasure to the user as a result of its function. For example, a shower unit sprays warm water over a user – an experience that can be sensually pleasing. Here, then, it is not the tactile properties of the product itself that are providing the pleasure, but rather the pleasure has accrued from the outcomes of the product's function. A similar analysis would apply to a product such as a foot-spa. Here, again, the product is designed to deliver a sensorial pleasure to the user – in this case a massaging of the feet – irrespective of whether the product has pleasing tactile properties in and of itself.

Another type of physio-pleasure that products can deliver as a result of their functions are pleasant tastes. For example, the taste of coffee produced by a coffee-maker may be dependent on a number of design issues, such as

Figure 2.3   Keyboard of the Apple Macintosh PowerBook

27

Figure 2.4   Pencil with a silicon–rubber
composite surround

the speed at which water drips through the filter and the temperature at which the coffee is kept after filtering. Indeed, many coffee-makers now have adjustable brewing processes in order to offer the user a variety of tastes.

Other food- and drink-related products that deliver physio-pleasures through their function include, for example, food processors, blenders, kettles and cooking apparatus. The basic issue with these products is to ensure that they support the user in preparing the food and beverage in the optimal manner. For example, a food processor should dice food to the appropriate size, and cooking apparatus should be made of materials that do not impart any unwanted flavourings to the food.

Another source of pleasures that a product can deliver to the user are olfactory pleasures. Perfumes and after-shaves, for example, can make the user smell pleasant and attractive. Some people also gain pleasure from the clean 'technical' smell associated with new products, such as new books or the smell inside a new car. Indeed, it is now possible to buy sprays that simulate the smell inside a new car. Used-car dealers often spray car interiors with this in order to make their vehicles more attractive to potential customers.

Other products can give the user physio-pleasure by altering the user's physiological state in a more general sense. For example, a product might help to make a person physiologically relaxed or physiologically alert. One design element known to have an effect on physiological arousal is colour. For example, the combination of red and black is thought to raise the viewer's level of physiological arousal. It is possibly for this reason that this

28

colour combination is often used on war propaganda and recruitment posters – by raising the reader's state of physiological arousal, the hope is that this energy will be channelled into negative feelings against the enemy.

Other colours have a more physiologically calming effect: pastel colours can lower a person's state of arousal, and the use of pastel colours might be appropriate, for example, in the interior of an aircraft. People often have to sit still for hours on end in an aircraft – it may be better that they are physiologically relaxed and perhaps able to doze off, rather than feeling restless for the whole flight. Of course, physiological arousal often parallels psychological arousal – again a reason to reduce arousal in a situation that many people find stressful.

Earlier, the dangers of products – such as chairs – inflicting physical damage on their users was discussed. However, it is also possible for products to bring positive health benefits to users. For example, products such as thermometers or blood pressure meters help people to monitor their health and to spot any early signs of potential problems, giving them the opportunity to seek medical attention before a major problem develops. Other health-monitoring products include blood sugar level monitors and medication dispensers.

Some products can bring direct physiological health benefits through their use. For example, fitness machines such as rowing machines or running treadmills can help the user to become physically fitter. Other products can bring benefits to the user through their effect on the chemical composition of the blood – a syringe is an example of a mechanism through which performance-enhancing or health-giving substances (including certain types of hormones and nutrients) can be injected into the bloodstream. (Syringes can also have negative associations for many, in particular because of their association with narcotics misuse – this will be discussed more in the next section.)

Socio-pleasures

Socio-pleasures are to do with – in the broadest sense – our relationships with others. Social need pleasures are those that enable us to be comfortable, or to avoid discomfort, in these relationships. Products that give need pleasures in this context are, then, those that help us, within particular contexts, to feel that we are socially accepted.

Consider, for example, clothing. Codes and conventions deem certain types of clothing acceptable in certain situations. Whilst jeans and T-shirt might be perfectly acceptable attire for going to the pub for a drink with a friend, it might be totally unsuitable for a board-room meeting at a very traditional firm of chartered accountants. By the same token a 'stuffy' three-piece suit might be looked on as odd attire for a night out at the disco. Wearing inappropriate clothing in a particular situation can cause the wearer embarrassment and create negative feelings towards the wearer in others.

Of course, the acceptability or otherwise of particular styles of dress will not only depend on the social setting, but also on the opinions and attitudes of observers. For example, younger people may expect informality of dress in situations where an older person would expect more formality. In many UK churches, for example, younger people may attend in informal leisure clothes, whilst older people may still feel that they should wear their 'Sunday best'.

Another example of a product whose social impact tends to be particularly context- and person-dependent is the mobile telephone. When mobile phones first came to the market owning one tended to be seen as a status symbol. The impression that mobile-phone users appeared to give was of being successful, dynamic and important – so important that they needed the mobile phone so that others could keep in touch with them at all times. There appear to many people who still hold this opinion of the mobile phone. A glance around many London pubs will often reveal a number of people proudly using mobile phones!

On the other hand, some people may regard those who use mobile phones with a degree of contempt – perhaps regarding the use of such devices as being vulgar pretension. Others may simply regard users of mobile phones as being ill-mannered and inconsiderate because of the disturbance that they cause to those around them. In his humorous and affectionate portrayal of Britain and the British, *Notes from a Small Island* (1995), American journalist Bill Bryson satirises a typical mobile-phone user, making totally unnecessary calls simply to show off the fact that he owns a mobile phone:

> 'Hello, love. I'm on the 10.07. Should be home by five. No reason to tell you at all except that I've got this phone...I'll call again from Doncaster for no reason...'
> '...Clive here. Yeah I'm still on the 10.07 but we had a points failure at Grantham so I'm looking now at an ETA of 13.02 rather than the forecast 13.00 hours.'
>
> (Bryson 1995: 187)

Mobile phones, indeed telephones generally, have a number of social issues associated with them. In Western society access to a telephone might be seen as a bare minimum level of social access – if a person has no access to a telephone they may feel very isolated. Perhaps, then, the basic function of person-to-person communication provided by a telephone might be seen as fulfilling a need pleasure, rather than a positive pleasure of appreciation. This is an example of how, over a period of time, a pleasure can move from being a pleasure of appreciation to being a need pleasure. In the early days of telephone-owning, the possibility of communicating over large distances may have been seen as something very special. Indeed, the inventor of the telephone, Alexander Graham Bell, once boasted that 'one day there will be a telephone in every major city in the USA' – an almost comical under-estimation of the profound effect his invention would have on society.

Sometimes a product may contribute to social nuisance. For example, products that make a lot of noise may not be appreciated by those in the vicinity. Many tools can be noisy and cause irritation for those within earshot – for example, many people will find it extremely irritating to have to listen to workers digging up the road with pneumatic drills. Similarly, hearing neighbours drilling or hammering can be very annoying, as those who have had the misfortune to live next door to 'Do-it-Yourself' enthusiasts will testify.

A curse on the typical British suburban Sunday afternoon can be the lawnmower. Many a potentially relaxing alfresco afternoon tea has been ruined by the sound of a neighbour cranking up their petrol-driven lawnmower, and the subsequent hour or so of thunderous roaring that everyone in the street has to put up with until the owner of the machine is satisfied that his or her lawn has been shorn to the appropriate level. Fortunately, more recent lawnmower designs are much quieter and less disturbing, and run, as they tend to be now, on electricity rather than petrol. The recent designs are, then, providing a need pleasure, by helping to eliminate a negative – the annoyance of excessive noise.

Another social curse, in the eyes – or rather ears – of many, is the noise pollution caused by other people listening to loud music: radios playing in the neighbours' garden; stereos booming in next door's living room; the hiss, clicks and crashes emanating from the personal stereo of the person sitting a couple of seats away on the train. All these can be very irritating to those exposed to them.

Psycho-acoustics provide a potentially effective solution to the problem of loud, booming bass tones that can resonate through the wall when a neighbour is playing the stereo very loudly. It has been discovered that when humans hear particular notes in a harmonic series the brain automatically completes the series – the effect is that the listener 'hears' notes that are not really there. This means that it is possible to give the user the effect of big, booming bass tones that are not really there, simply by emphasising other notes in the harmonic series related to the bass notes. This technology is currently used in some of the effects boxes used by professional musicians. In the context of household stereo systems, application of this principle enables the listener to enjoy booming bass tones without the disadvantage of having sound resonating through the walls or floor and disturbing others.

Noise is not the only source of social irritation or annoyance that products can create. Products can also be visually disturbing. For example, whilst a product such as a garden gnome might be seen as fun and cheerful by the owner, others living in the same street, who see it every time that they look out of their windows, might find it gaudy and tasteless. They may think that people who put gnomes in their gardens are bringing down the whole 'tone' of the neighbourhood! A 'tastefully' laid-out garden, on the other hand, is likely to win the neighbours' blessing.

Of course, the appearance and placing of the buildings themselves can have a large effect on whether a neighbourhood is a pleasant or unpleasant place to live. This has been an issue that has attracted a lot of attention in the UK recently due to the outspoken views of Prince Charles, the heir to the throne, with respect to some of the architecture in Britain. The prince famously described a new building in London as a 'monstrous carbuncle' on the face of the city. He has also worked with a group of architects to design what he regards as an architecturally ideal village, which has subsequently been built in Cornwall, in the south-west of England. The idea behind this is that new buildings should blend in visually with those that are already there, so as not to create disharmony in the visual landscape. It could be argued – depending on one's taste – that this approach is fulfilling a social need pleasure by creating a visual environment that will not offend.

Another social role played by architecture is in the discouragement, or unwitting promotion, of crime. Many of the housing estates built in the 1960s and 1970s have been designed in a manner that seems to encourage crime of one sort or another. Designs that force people to walk through dimly lit passages in order to get to their dwellings will inevitably make it easier for the unscrupulous to assault or mug these people. Similarly, darkly lit corners on such estates make for ideal places for drug crime to thrive, both in terms of the selling and the taking of drugs.

Other, more subtle, aspects of the layout of such estates have been shown to have a significant effect on the level of vandalism that can occur. For example, the design of the entrances to and walls of tower blocks and the immediate surroundings of the tower block can either encourage or discourage vandalism. It seems that the major factor is the degree to which the design of these aspects communicates 'ownership' of the flats. For example, if the tower block has just one main ground-level entrance, serving the whole of the building, then this has the effect of appearing to make each individual dwelling seem more anonymous. On the other hand, if there are a number of main ground-level entrances, this gives the effect of making each dwelling seem part of a little section of dwellings, rather than as simply part of the tower block as a whole. If these entrances protrude away from the building, they have the effect of seeming to set the dwelling apart from the communal ground around it – this can contribute to an image that the block consists of a series of private dwellings, rather than simply being a public building set in a strip of public ground. It has been shown that these sorts of design approaches will reduce the level of vandalism against such property. Whilst the 'hardcore' vandal may not be deterred, the more 'casual' vandal may feel uncomfortable at the prospect of defacing a building that he or she feels has 'owners'.

It may also be possible to design products for public use in a manner that discourages their vandalism. For example, it may be that some materials are more attractive to vandals than are others. The shattering of glass, for example, may be more attractive to a vandal than the cracking of Perspex. For this reason, Perspex might be a suitable material for use in, for instance, a telephone kiosk.

Another social problem associated with products is theft. A number of products are designed to discourage their own theft, whilst other products are designed to prevent the theft of something else.

The car stereo provides an interesting example of the first of these types of product. As car stereos became more and more elaborate and appealing, they became more and more of a target for thieves. By the mid-1980s car stereo theft had become a very prevalent crime, to the extent that people started to question whether it was even worth having a car stereo. Not only was it possible that the stereo would get stolen, but there was the added problem of the car itself being vandalised during the theft – typically the thieves would smash the glass in the driver's or passenger's door, in order to be able to gain access to the car in the first place.

Car stereo designers have taken a number of different approaches to solving this problem. One solution was to make the stereo inoperable without the user first entering an alpha-numeric code. As long as the thief didn't get hold of the code then he or she, or a third party to whom the product might subsequently be sold, would not be able to use the stereo – there was, in theory, then, no motivation for anyone to steal the stereo.

The problem with such an approach was that the thief would probably not know that he or she could not use the stereo until after getting it home and trying to use it. In fact, if he or she were to sell the product on, then it might only be the subsequent purchaser who would suffer the disappointment of finding the product inoperable. This design feature, then, did little to discourage the theft in the first place – the owner was still left with a vandalised car and without a car stereo.

Another approach to addressing this problem was the car stereo with the removable front panel. In this solution, the controls to the stereo and some essential electrical connections that are required for the stereo's operation are placed on a single removable panel. When leaving the car, the driver can remove the panel and take it with him or her. If a potential thief came and looked inside the car, he or she could clearly see that the front panel had been removed and would realise that there was no point in breaking into the vehicle – the owner, then, would not have a vandalised car and would still have a car stereo. This, then, seems a far better solution to the problem.

Another social need pleasure may be the desire to avoid stigmatisation – a form of social labelling that carries negative connotations. This issue was recognised by a company that manufactures a device known as NovoPen[tm]. This product is for use by diabetics for injecting insulin. It is designed to look like a pen and can be clipped into a jacket or trouser pocket. The pen-like appearance of this product contrasts sharply with that of a syringe. Syringes can have negative social connotations. First, their medical associations may emphasise the person's condition, something which he or she may not necessarily want to communicate to others. Second, their associations with intravenous narcotics use is something that can be stigmatising. By drawing on the metaphor of a pen, the NovoPen[tm] plays down the medical nature of the person's condition and avoids all narcotics associations. The NovoPen[tm] is illustrated in Figure 2.5.

Aside from helping to avoid social pitfalls, products can also contribute to creating positive social consequences – both for individuals and for society as a whole.

Above, it was noted that clothing can have an effect on how people are perceived in particular situations. In particular it was asserted that the manner in which a person is clothed may have an effect on the degree to which they 'fit in' in a particular situation. However, clothing can do much more than simply help in the avoidance of social *faux-pas* – it can help a person to establish a positive, affirming social identity.

One way in which it can do this is to identify the wearer as a member of a specific social group. For example, Dr Marten's boots have become an item associated with skinheads – wearing these boots is part of what identifies someone as being a proper skinhead as opposed to a person who simply happens to have very closely cropped hair. Wearing the boots, then, can contribute to a sense of belonging for those who wish to be part of such a group. Baggy jeans and T-shirts play an equivalent role in rap culture.

Figure 2.5   The NovoPen[tm]

Of course, it is not only clothes that send out social signals in this way. In the late 1960s and early 1970s the Volkswagen camper van became strongly associated with the hippie movement. Not every hippie could afford one, but if a person did own one – particularly one painted in bright psychedelic colours – it was a sign of belonging to a particular group. Indeed, the purpose and function of the vehicle – and its associations with freedom and the open road – fitted well with the hippie ideals. It was also a vehicle that encouraged sociability, enabling a number of people to travel together, camp together and eat and drink together. It became one of the icons of the hippie culture.

The clothes that a person wears may often be a clear signal of the institution or community that a person belongs to. For example, a priest's 'dog collar' clearly identifies him or her as a member of the clergy and as a Christian. It is giving a clear signal to society about who he or she is and what he or she believes. In certain situations or communities – for example in church or in particular communities – wearing the collar will also give him or her authority over others

A very clear example of how clothing can reflect a person's authority is the military uniform. Here the wearer's rank is clear from the insignia that he or she wears. The uniform, then, is integral to the wearer's authority. If a five-star general were to walk into a base where he or she was not recognised, then the way that the soldiers on the base react to this person is likely to be highly dependent on whether he or she was in uniform.

Closely allied to authority is status. Bourdieu (1979) identifies two different types of status that products can convey on their owners and users – material status and cultural status. Material status is conferred by products that give an impression – rightly or wrongly – that the owner has significant material wealth. Objects that confer cultural status are those that give the impression – again rightly or wrongly – that someone is a person of great cultural knowledge and taste.

In the yuppie era of 1980s Britain, perhaps the ultimate status symbol was the Porsche sports car. This vehicle became synonymous with the young successful executive, who worked in financial institutions and took home a huge salary. Owning a Porsche made an implicit statement: 'I'm young, I'm rich, I'm successful.'

Whilst there is no suggestion that owning a Porsche is a sign of 'vulgarity', those who openly flaunted their wealth during this big boom in the City were often looked down on by others who felt that they flaunted their wealth in a vulgar and crass manner. In his television show during this period, British comedian Harry Enfield satirised the yuppie culture through a character called 'Loadsamoney'. This character, adorned in chunky gold chains, rings and bracelets, would tell everyone he met how rich he was and would always be carrying a substantial wad of bank-notes with him.

35

Whenever he went into a shop he would 'wap his wad on the counter' before making a purchase, even if it was just a newspaper or a packet of cigarettes.

Characters such as 'Loadsamoney' are an exaggerated version of what Bourdieu (1979) would characterise as a person with a high material status but with a low cultural status. As well as his chunky jewellery, he is probably the sort of person who would enjoy 'flashy' products: stereos with hundreds of flashing lights; and designer-label clothes mixed and matched in a random, tasteless sort of way – for example, the best Nike sports shoes worn with an Armani suit. He would drink expensive German lagers straight from the bottle so that his fellow drinkers could see that he was drinking an exclusive brand and he would eat in expensive foreign restaurants whose menus he would not be able to understand.

'Loadsamoney' represents the sort of character that those who value cultural status might look down upon. For these people it is far more important to consume products that demonstrate taste and cultural appreciation. In terms of styling, these people are likely to prefer understated design language as opposed to flashiness. Rather than aspiring to own the latest German sports car, they might go for a classic British model, such as an MG. They may enjoy products made from 'noble' (natural, high-quality) materials, such as metals and woods. They may have a few carefully chosen antiques in their house, rather than the latest offering from the Milanese studios. They might drink wine rather than lager.

Of course, the description in the above paragraph is as much a parody of the cultural-status seeker as 'Loadsamoney' is of the material-status seeker. The intention of the caricature is not to accurately identify a particular target group, but simply to demonstrate that products can confer status to different people in different ways.

Modern production processes have made it increasingly feasible to manufacture products in small batches. This means that it is now possible to create and manufacture small quantities of a particular product aimed at a limited target group and, despite the low production volume, to make these products affordable. Such products may appeal to cultural-status seekers as they enable them to choose something exclusive that people who buy 'mainstream' products would not appreciate. Consider, for example, the citrus press, designed by Philippe Starke for Alessi (see Figure 2.6). The unique styling of this product might make it appealing to those wishing to make a cultural statement. Indeed, owning a 'Starke' may be seen by some as a way of buying cultural status. If a person simply wanted to squeeze a lemon he or she would probably be better off going to Woolworths and buying a cheap plastic juicer for fifty pence.

The issues addressed in this section so far have focused on how a person's product choices can affect how he or she is perceived by society as a whole. Another issue that falls under the heading of socio-pleasure is the role that a

Figure 2.6 Juicer designed by Philippe Starke for Alessi

product plays in the relationship between a person and the individuals with whom he or she is close.

Some products have particular significance as gifts. For example, giving another person an item of jewellery is often an expression of love and affection. Indeed, in Western society, the wedding ring is perhaps the strongest token of love between a man and a woman. Other products are used as part of processes that can be expressions of love and friendship towards others. For example, for British people, the basic courtesy shown to friends when they pop round to visit is to make a cup of tea for them. The sight of the teapot on the table might almost be seen as the sign of a friendly gathering.

Cooking is also a process that can be loaded with social significance. Again, people might see cooking for others as a token of affection towards friends and loved ones. Cooking a meal for a partner may be an element in a romantic evening. This, in turn, means that the products used in this activity can take on a social significance. For example, serving guests using attractive crockery may be seen as a sign of respect that contributes to making a meal into a special occasion. On a more practical level, efficient and effective kitchen equipment helps to make sure that the preparation of the meal runs smoothly and that a high-quality meal is provided.

Furniture is a product area that is often heavily loaded with social significance. A look around many offices will reveal that the size of a person's desk can be a fairly accurate indicator of that person's status within the organisation. Certainly, there seems to be a perception that the bigger the desk, the more powerful the person sitting behind it is. Indeed, this can sometimes be intimidating for the person sitting on the other side.

One type of furniture that has encouraged designers to be playful as regards the product's social role is seating. For example, the lovers' seat – a sofa designed such that those using it sit face to face – forces those using it into close facial proximity! Even more straightforward furniture can play a significant social role in a family situation. The stereotype nuclear family – where dad sits in his armchair puffing his pipe, mum sits in the corner knitting and the kids squash up on the sofa – was once seen by some as being indicative of the ideal of domestic bliss. This may seem a rather dated view, but the sofa still retains a major social role in the household, whether it be the single person curled up with a magazine, a group of young men drinking beer and watching the football, or two young lovers kissing, the sofa can be a 'social centre' within the household.

Other products derive their social significance, not only though reflecting social issues, but also by shaping the society in which they are placed. Perhaps the most obvious example in this generation is the computer. This has changed the nature of certain aspects of social communication. For example, e-mail has become a tool for leaving casual messages for others in the same way that people may previously have left a note on a fridge door for a house-mate. The difference being, of course, that with e-mail it is possible for a Londoner to leave a note for a New Yorker as easily as sending a message to the person in the office next door. An etiquette for e-mail communication has developed that allows for brief, informal, to-the-point communication in the way that previous media have not supported. This applies both to formal and to informal communication.

For example, it seems perfectly acceptable to drop a couple of lines to a friend using e-mail, whereas people would probably expect a more lengthy correspondence in a letter or a more substantial discussion on the telephone. Similarly, with work issues, it is common for colleagues to send messages that are shorter and more informal than would be put into a more formal memo or fax message. Whilst most people would probably regard this development as a good thing, it has been shown to have its dangers. In particular, the informal atmosphere surrounding this communication medium has often led to colleagues using the e-mail to discuss issues that they would probably have regarded as far too confidential for correspondence via fax or memo. Recently, particularly in the USA, there have been a spate of legal cases in which lawyers have demanded access to the e-mail archives of companies in the hope of finding evidence to incriminate or exonerate defendants. Often they have been successful in their search. This has led to many major firms developing strict guidelines for how their employees should use the e-mail and about the sorts of topics that they can and cannot discuss.

The Internet has also been at the heart of something of a social revolution. One way in which this has been so is in the increased freedom of expression and of information. This has made it increasingly difficult for governments to suppress information from their population. This was illustrated starkly in the UK in 1998. In that year the son of a senior British government minister was accused by a journalist of buying illegal narcotics. The British courts imposed an injunction on the British media, forbidding them from giving any information about which minister's son was involved. This resulted in a blanket news ban on divulging this fact. However, the courts had no effective way of censoring the Internet. Those Britons who had access to the Internet could find out who was involved straight away.

Some may regard the example given as trivial within itself, but it is indicative of a wider principle. Many would see citizens' rights to be informed and to be heard as basic human rights. It could be argued, then, that Internet users have, in effect, more opportunity to exercise their human rights than those who are non-users. It may be argued that giving Internet users these rights is not always good for society as a whole – some, for example, choose to express themselves by giving information about how to make bombs or by loading the Internet with pornography. Nevertheless, from the point of view of those expressing themselves, or those accessing information, the Internet is providing a social pleasure.

Another way in which the net has brought socio-pleasure is through creating what is effectively an alternative social space – Cyberspace – in which everything is 'local'. It is as quick and easy to access information about a book-store in another country as about the book-store in the next town – and just as easy to order a book. The Internet chat rooms make it just as easy to discuss a favourite topic with someone on the other side of the world as with someone in the local pub. Of course, there are many ways in which 'real' world and Cyber-world experiences differ, but, at minimum, the Internet provides a means by which individuals can, at least in terms of information exchange, break free from the constraints of their local social environment.

Of course, it is not only through providing a platform for e-mail and the Internet that computers have affected society. Through providing a vehicle for the automation of many tasks and jobs, they have had a large social impact on the workplace. A trend that is starting to emerge is an increase in the number of professionals working from home. To many people – in particular mothers with young children – this has provided the opportunity to raise children with comparatively little hindrance to their careers. This is in contrast to the traditional workplace where having to spend working hours away from home may mean a choice of children or career. Even as we embark on a new millennium many companies still seem to have discouraging attitudes towards those who want time off to raise children and many more are without child-care facilities.

Despite the increasing uptake of Internet use, for most people – certainly most Western people – the main 'window on the world' is still the television. It is through the television that many are kept informed of what is going on in the world and about decisions that are being made by politicians that may affect their lives. Again, receiving this information may be seen as a basic human right and is part of what makes a person an integrated member of a society. However, it is not only in providing factual information about a person's environment that the television promotes social inclusion. Because watching television is, in the West, an almost universal pastime, other types of programme – such as sport, entertainment and drama – can provide common talking points for people in a society.

For example, nearly all English football fans of this generation will remember where they were on that infamous night in July 1990 when England lost the semi-final of the World Cup to Germany on a penalty shoot-out. They will also remember what they were doing – watching television. Television, then, has created a bond between these people, by offering them a shared experience. English football supporters can talk about this game as a shared experience, even though, bar the few thousand English who were able to attend the game in Turin, none of them were there. By facilitating the creation of a common talking point, television can help in creating bonds between people, perhaps even total strangers. This may be part of what gives people a shared social identity and a sense of belonging.

As with football, soap opera can provide a common frame of reference that helps to facilitate social interaction. In British workplaces, soaps such as *EastEnders* and *Coronation Street* are the subject of much conversation, with many enjoying gossiping about the characters and giving their opinions on what they have seen in last night's episode. Again, being part of this banter helps establish a person in a social group and can give a sense of belonging and social identity. In the late 1970s and early 1980s, much of Britain became gripped with the events unfolding in the glamorous American soap series, *Dallas*. This series followed the fortunes of the Ewing family – Texan oil tycoons. Events in the soap reached a climax when leading character J.R. Ewing was shot. Record British audiences tuned in for that episode, and trying to work out who had shot him became a national pastime over the ensuing weeks. Indeed, a whole cottage industry appeared to spring up around this episode. Stickers and badges saying 'I shot J.R.' and 'I hate J.R.' could be seen adorning a host of cars, jackets and bedroom windows. Meanwhile, the tabloid press dedicated acres of space to analysis and speculation as to who the would-be assassin was. Even bookmakers got in on the act – it was possible to place bets on who the culprit may have been.

This, like the semi-final of the 1990 World Cup, was a shared national event, facilitated through the power of television. Of course, there are many people who do not like football or soap opera, but, for those who do enjoy them, the sharing of the experience with others, even people they do not know, can be part of the fun.

As a final example in this section, consider the socio-pleasures facilitated by a number of different forms of transport. Like telecommunications, transport has a major role to play in terms of broadening people's social horizons. Indeed, it may be argued that it is even more salient. After all, whilst telecommunications products such as the telephone, the mobile phone, e-mail and the fax may be used as a means of maintaining a friendship or an intimate relationship, these are rarely the media through which the relationships are initially formed. Friendships and intimate relationships tend to be forged in face-to-face situations. Transport has played a major role in expanding the opportunities to meet people outside of the local surrounds.

In the days before the bicycle, people's social circles may have been confined to their local town or village. The friendships and relationships that people forged may have tended to be with others that lived in the same town. However, with the invention of the bicycle, this expanded significantly. Journeys that may have been impractical on foot now became easier to make. People may now have been more inclined to go to neighbouring towns for a night out – for example to dances or other social events. Now, friendships and relationships across town boundaries became more common. Here, then, the bicycle provided people with a socio-pleasure – the expansion of their social horizons.

As car ownership has become more and more common and as travel by train and bus has become widely available and affordable, social circles have expanded further. People may now come from miles around to attend a communal social event, such as a concert or a rave, meeting up with others from towns and cities in different parts of the country. The availability and affordability of air travel has taken this a significant stage further. Now people from all over the world may meet up in popular holiday destinations.

Psycho-pleasure

Usability is a product property that might be seen as being associated with psycho-pleasure. Products may be difficult to use if their use puts too high a cognitive demand on the user. This, in turn, may result in unpleasant emotional responses such as annoyance, frustration or stress. Usable products might be seen as fulfilling a need pleasure through helping the user to avoid such unpleasant feelings

As discussed in Chapter 1, a product that was seen as being revolutionary with respect to usability was the Apple Macintosh personal computer. The first widely available computer to come with an easy to use direct manipulation interface, this product 'demystified' computers for many. Whilst earlier personal computers, with their command line interfaces, may have seemed unapproachable, threatening and difficult to use, the Macintosh empowered users by drawing on metaphor to enable the person with little or no experience or knowledge of computers to start using the latest technology.

Non-usable products frustrate because their interfaces are, in effect, providing a barrier between the user and the technology and functionality that the product has to offer. With usable products, by contrast, the interface might be seen as a gateway to the functions and technology, empowering the users and giving them feelings of confidence and control.

The Macintosh is easy to use partly because it hides the technology from the users. The design disguises the technology in a desktop metaphor that allows the user to think in terms of 'dragging objects', 'selecting menus', 'opening folders' and 'throwing things in the trash' rather than in terms of 'executing command strings' or 'processing routines'. An assumption behind this – one that may be valid for a great many cases – is that people are not interested in the underlying technology, but simply in what the product can do for them.

Facilitating the completion of unpleasant tasks in as 'painless' a manner as possible can also be seen as a psycho-need pleasure. Domestic chores, such as vacuum cleaning and ironing, might be seen by many as being unpleasant. In these cases, products that help the user to get the job done as quickly as possible might be seen as fulfilling a psycho-need pleasure. In the case of irons, for example, the material used for the surface of the ironing plate can be an issue that affects the speed with which the iron can be moved across the clothes. This is a characteristic of the iron that is likely to have a real effect on performance. When customers are asked why they chose one iron over another, the perceived speed of the iron is often quoted as the rationale. Interestingly, though, the criterion by which people most often judge the perceived speed of the iron tends not to be the material of the ironing plate, but rather the shape of the iron.

Irons that are sharply pointed at the front and that are lower in terms of overall product height, tend to be perceived as faster. Strange as it may seem, it appears that people tend to project a speed metaphor on to irons that is derived from sea-going vessels. In this metaphor, the front of the iron represents the keel – the sharper the keel, the faster a boat will move through the water; the sharper the point of the iron, the faster people perceive that the iron will move across the clothes. The lower a boat, the less wind resistance it will meet as it moves along; again, the lower the iron, the faster people perceive that it can be moved. In other words, the closer the shape of an iron to the shape of a speedboat, the faster people are likely to perceive it as being. This association seems almost certainly subconscious – after all, when they thought about it, most people would surely not expect that these factors would have any real influence on the speed at which the iron can be moved. Indeed, sharpening the angle of the ironing plate can, if anything, make ironing slower as it will cause an overall reduction in the surface area of the ironing plate, reducing the amount of cloth that can be smoothed at a time. Nevertheless, the metaphor is one that is well known amongst iron designers and one that has long been employed within the industry.

In the case of vacuum cleaners manufacturers have employed three main strategies in order to make the task faster and more convenient. One of these is to increase the suction power associated with the product. Essentially, this is a technical issue, largely related to the power of the motor within the cleaner. However, the designer may have a role in illustrating and accentuating the perceived power of the cleaner. One way in which this can be done is through the form of the cleaner. For example, the shape of the vacuum cleaner may have strategically placed bulges, giving the impression of a vast, powerful motor that can barely be contained within the housing. This might be seen as a reflection of the design of powerful hot-rods, where the top of the engine – literally, in the case of some hot-rods – sticks through the top of the bonnet. Similarly, the housing may contain slots and grills, again drawing on the metaphor of powerful motor vehicles.

In the case of products such as vacuum cleaners, where power is strongly linked with quality of performance, the sound of the product may have a major role to play in convincing people of its quality. Vacuum cleaners have tended to be annoyingly noisy products: both the sound of the motor and the sound of the air moving through the product causes disturbance to the user and those in the vicinity. Because of this, manufacturers have tended to try to reduce the noise associated with these products. Indeed, up until the early 1990s, quietness was seen as a major selling point when people made a purchase decision for a particular cleaner. However, acoustic technology has now advanced to such an extent that it is now possible to produce powerful vacuum cleaners that are, if not entirely silent, certainly very quiet. Unfortunately for manufacturers making such products, many people find it difficult to believe that vacuum cleaners that are so quiet can actually give a good level of performance. As a result a number of manufacturers reversed their search for silence, deliberately making products noisier than necessary in order that they communicated a sense of power.

Another example of a product where manufacturers have added noise in order to give an impression of enhanced performance is the video cassette recorder. These products rely, for their operation, on a number of rather cumbersome mechanical movements. The sound of these can give a rather low-tech impression – the product can sound slow and lumbering as the levers and components clatter around. Slowness can be irritating, especially when the user is rewinding or fast-forwarding the video cassette tape. Because of people's irritation with this slowness, a number of manufacturers have added sound chips to their machines. When the machine is in rewind or fast-forward mode these chips play a sound sequence imitating a winding mechanism accelerating up to a huge speed, giving the impression that the tape is being wound very quickly. In reality, of course, the tape is not being wound any faster than it would be on any other equivalent machine – the added sound is simply giving the impression of enhanced speed.

Another case in which manufacturers' striving towards silent operation was reversed is that of the dishwasher. Again, as with vacuum cleaners, quietness in operation had become a major competitive issue amongst dishwasher manufacturers and, as with the vacuum cleaners, advances in acoustic technology had, by the early 1990s, made it possible for a dishwasher to operate in virtual silence. Unfortunately, because these machines were so silent, people often thought that they were not working at all. Because the cleaning goes on behind a closed door, people cannot see that the product is working and sound is, perhaps, the major cue that makes it clear that something is happening inside the machine. This might seem a trivial issue – okay, so people may be concerned the first time that they use the product, but after that, when they see how well the product cleans their crockery and cutlery, surely the silence would be seen as a great bonus?

Unfortunately, for a significant number of customers, their machines were broken before the first wash cycle was ever completed. The reason for this is that, because they thought that the machine was not working, they tried to open the door to check if there was a problem inside the machine. Dishwashers are designed with a door lock to prevent people from opening the machine whilst it is in operation. Unfortunately, however, some people, perhaps in their anger and frustration at a new appliance that was apparently not working, pulled so hard at the doors that they broke them off the fronts of the machines, ruining the dishwashers and flooding their kitchens with water. Here, then, is an example of a product design that failed to reassure its users that the product was functioning correctly, with the result that frustrated customers ended up breaking the product.

The time period just after buying a new product can be important psychologically. During this period, particularly if the product is an expensive purchase, people may want to feel reassured that they have made the correct purchase choice. Indeed, people will try to find evidence to reassure themselves. Evidence for this comes in the form of a survey of people who read car advertisements. Apparently, people are more likely to read the advertisements for their new car after they have bought it than they are during the decision-making process (Soloman 1996). This phenomenon is known as cognitive dissonance (Banyard and Hayes 1994) – the search for evidence that confirms what a person wishes to believe. Clearly, people are likely to want to believe that they have made sensible or appropriate purchase choices, especially where a high expenditure is involved.

Clearly, advertisements are one source of positive information about the product; however, the design of the product itself can also send positive, reassuring messages. One issue for manufacturers to consider here is the first things that people will do with a product when they get it home. Do parts of the product need to be assembled? Perhaps they should fit together with a positive sounding click. Do people have to put batteries into the product? Perhaps the lid to the battery compartment should feel sturdy and well-constructed, reassuring people through its tactile properties that they have bought a solid, reliable product. What about the dials and switches? Again, if their tactile properties communicate a solid, reassuring feeling people are likely to feel reassured about product quality.

However, it is not only the issue of build quality that may concern people at this early stage. Perhaps they will also be concerned about how easy the product is to use – they may want to get started quickly, to see the product working and get to grips with its main functionality as quickly as possible. In this sense, guessability – the ease of first-time use – becomes an important issue. This issue has come to the attention of a number of manufacturers of software-based products. For example, a number of computer packages now include on-line tutorials to enable the user to quickly familiarise himself or herself with what the product can do. Similarly, software-based games, such as those manufactured for the Sony Playstation and the Sega Megadrive, often include quick-start modes, enabling the user to have a go at the game without having to learn the detail of the interaction design. These early experiences can serve to reassure people that the software works well and that it will be useful or entertaining, providing an incentive to go on and learn about the full potential of its more advanced features.

Games are an example of a product type that are designed primarily to promote emotional enjoyment through providing people with a cognitive and physical challenge. In this sense, they provide the user with psycho-pleasures of appreciation. As an example, consider the simulation soccer game, *FIFA '99*, designed by EA Sports for the Sony PlayStation. Here, the player selects a team that he or she will control. This team is then pitted against others controlled by the computer or by other players. Success at the game requires a combination of cognitive and physical skill. The player must make decisions about tactics and must be able to make rapid decisions about what the best option is in any situation during the play. He or she then has to use physical skill to execute the intended action via the controls on the joypad.

Well-designed games can engage players in what they are doing. Instead of having the feeling that they are sitting in front of the television control-ling animated sprites via a control pad, they may feel that they are playing soccer at Wembley Stadium or trying to escape from a monster in some fantasy world. The concept of engagement was first identified centuries ago by Aristotle when formulating his theory of poetics. Aristotle claimed that a measure of quality by which a work of fiction could be judged was the extent to which the audience became engaged by the story. For example, a theatre audience should, in a sense, 'believe' in what is happening on stage, rather than simply thinking of the work as a group of actors moving around the stage and delivering lines. In the case of computer games, issues such as the quality of the graphics, sound and game-play can have a major role to play in determining how engaging or otherwise a game is. Game-play basically refers to the interaction design of a game – the extent to which the options available to the player and the actions required to play reflect the real-life or fantasy situation simulated by the game.

Laurel (1991) has taken Aristotle's idea and applied it to computers, claiming that a sense of engagement with the 'world' of the program that the person is using can be a central factor in determining whether the user experiences a positive joy. This has been cited as an example of the benefits of using metaphor in a computer interface. For example, Clarke et al. (1995) report the development of an information retrieval system built around the metaphor of a library. Clarke et al. developed an interface that included virtual agents representing, for example, librarians and experts with respect to the material being viewed. The user interacted with the system by 'communicating' with these agents who would then carry out tasks, such as information retrieval and the cross-referencing of material, on the user's behalf. When compared, in an empirical evaluation, to the original menu-driven interface for the same system, Clarke et al.'s system was unanimously rated as being both the more engaging and the more pleasurable to use.

An issue to consider when developing a new product is the nature of the emotion that it is desirable to elicit in those who will experience the product. For example, do people want to feel relaxed when using a product or do they want to feel excited? What is it about the design of the product that contributes to eliciting a particular response?

It has been argued that emotional responses to products can be linked to product aesthetics, for example to their form and colour. Macdonald (1999) asserts that people's responses to a particular aesthetic will be dependent, in part, on their cultural values and, in part, on their natural human instincts. In the case of colour, for example, there are some colours that humans may instinctively perceive as 'safe' and therefore relaxing, and others that might be perceived as 'dangerous' and which therefore may elicit emotions such as excitement or fear.

A colour that may be associated with safety and relaxation is blue. It may be argued that, because of its association with life-giving water, humans are instinctively drawn to blue and will feel relaxed when surrounded with blue. Red, by contrast, may be associated with danger. It is a colour associated with markings on a number of poisonous animals as well as with blood.

Because of these associations people may become more stimulated when they see red colouring. This may account for its usage in products associated with safety-critical situations – fire extinguishers, emergency shut-off buttons, stop lights and warning signals seem to be red the world over. However, because it raises people's levels of alertness and stimulation, red can also be associated with excitement. This might account for the extent of its use as a colour for sports cars, probably the most famous example being the red used by Ferrari on almost all of its models.

Car manufacturers also draw on people's natural aesthetic responses in order to give their cars 'personalities', which will contribute to eliciting particular responses. The front of vehicles can look rather like faces – the lights as the eyes, the bumper and the radiator grill as the mouth. Some cars may look aggressive, powerful and exciting; others calm and safe. The body-work may also mirror animal, or indeed human, characteristics. A car might look 'lean', 'perky' or 'muscular'. Macdonald claims that the form of some four-wheel-drive jeeps makes them look as if they have been 'working out in the gym' (Macdonald 1999: 98).

The sort of emotional response that is desirable in any given situation may depend upon the context in which a product is used. For example, some products are used in contexts in which the person may already be highly stimulated, whilst others may be used in situations where a person may be relaxed or even bored. Sometimes a product may provide pleasure through accentuating the emotion already present, whilst sometimes the pleasure may come from balancing that emotion. As an example of the latter case, consider the design of aircraft cabins. Flying can be an exciting experience – for some a frightening experience. Unfortunately, when flying, the passengers have to sit in a fairly confined area, perhaps for many hours, depending on the length of the flight. Having to sit still for a long period of time when feeling excited, or particularly when feeling frightened, can be frustrating. With this in mind, the colour schemes used in cabin interiors are typically chosen to reduce the passengers' levels of stimulation. Blue – a relaxing colour – is one that is often used, as are pastel colours. It would be unusual to see an aircraft cabin decorated in stimulating colours, such as bright red.

Products can provide psycho-pleasures of appreciation not only through their emotional benefits, but also through extending or enhancing people's cognitive capabilities. Norman (1988) refers to this as the 'concept of knowledge in the world'. He draws a distinction between the knowledge that people store 'in their heads' and knowledge that is stored in external devices – 'knowledge in the world'. An example of a product that stores knowledge is the electronic personal organiser. This can be used for storing information pertaining to a person's agenda as well as other useful information, such as telephone numbers, and train and flight schedules. Some personal organisers also include modems and GSM phone connections, giving e-mail and Internet access – all this in a device small enough to fit into a jacket pocket.

Clearly, trying to keep all of this information in the head is likely be impossible – whilst it may be possible to memorise a few appointments and telephone numbers, it certainly wouldn't be possible to memorise the entire contents of the Internet! Carrying a personal organiser can, then, give a person a readily accessible source of knowledge and information, far in excess of what he or she could store in the head.

Ideo-pleasure

Ideo-pleasure pertains to people's values. These includes, for example, tastes, moral values and personal aspirations. The issues that fall under ideological pleasure are, then, important in defining how people do and would like to see themselves.

One way in which people can differ markedly is in their attitudes towards modern technology. Whilst some may enthusiastically embrace technology and enjoy seeing technology around them, others may enjoy the benefits of technology, but may not enjoy technology *per se*. People's attitudes with respect to technology may influence the types of design aesthetic that they are likely to appreciate. People with positive feelings towards technology may prefer aesthetics that emphasise the technological aspects of a product, whilst those with negative feelings towards technology may prefer aesthetics that play down the technological aspects.

An example of a design aesthetic that emphasises the technological aspects of a product can be seen in Figure 2.7. This stereo system, manufac- tured by Sony, contains technical-looking displays containing digital readouts and flashing lights, and also makes use of futuristic form language. The use of metal, and the metallic colour scheme on the plastic body of the product, also give it a technological feel. The design of this stereo contrasts markedly with that of the Bang and Olufsen stereo illustrated in Figure 2.8. Here, the design aesthetic is comparatively 'quiet'. The stereo is designed to blend into its surroundings – almost to become a part of the 'furniture'. The technological aspects of the product are not apparent in the form language of the product, but rather are hidden beneath the product's exterior.

Some designs may be successful because they reflect the values of a particular era – because they tap into the spirit of the age: what is called in German the '*Zeitgeist*'. In her study of American furniture design in the 1950s, Greenberg (1995) describes the success of the 'jet age aesthetic' (Greenberg 1995: 14) . This aesthetic was reflected in the design of products with flowing lines and sweeping curves. Aside from the furniture designers that drew on this aesthetic, perhaps the most well-known exponent of this sort of design was Raymond Lowey. Lowey created designs for all sorts of products – from locomotives to refrigerators. The common element in the styling of these products was their streamlined aerodynamic form. In some cases, such as with locomotive design, it could be argued that there may have

Figure 2.7   Sony stereo

Figure 2.8   Bang and Olufsen stereo

been an element of rationality to such an approach. After all, locomotives are kinetic products and streamlining may be a performance-enhancing of their design. However, for static products such as fridges and other household appliances, the appeal was purely emotional.

Votolato (1998) explains the popularity of jet-age aesthetics in 1950s America as reflecting a desire amongst Americans to look to the future. The country had just emerged from two troubled decades, during which the American people had experienced first the Great Depression and then the Second World War, yet by the beginning of the 1950s America had emerged as probably the most powerful and prosperous nation in the world. After suffering for so long, Americans were looking forward to a brighter, wealthier future and associated the future with optimism and progress. Futuristic, jet-age styling was evocative of this brighter future.

The UK took longer to recover from the war than did the USA. Some types of food and other types of essential provisions were still in short supply during the 1950s and rationing remained in place throughout much of the decade. Nevertheless, things were improving – the post-war governments had laid the foundations of a Welfare State, providing a social safety net for those in financial hardship and providing all UK citizens with free health care via the newly founded National Health Service. However, the country's full recovery did not become apparent until the 1960s. It was in this decade that many people in the country began to look forward. After half a century that had brought two devastating wars and the collapse of much of the British Empire, there seemed to be a feeling that it was time to leave the past behind and to move on to a brighter future.

By this time, the jet age had turned into the space age and it was the aesthetic of the latter that characterised the optimistic British design during this decade. The use of rounded forms – reflections of the aesthetic associated with spacecraft and space clothing – was a result of this trend. The trend was also reflected in the use of shimmery, shiny materials in the fashion industry – these were reminiscent of the materials used for spacecraft and space clothing.

Aside from trends that change rapidly over time, countries may have ingrained cultural characteristics and values that will influence the extent to which certain design characteristics are appreciated. In the late 1960s and early 1970s, Dutch anthropologist Geert Hofstede conducted a survey in over fifty countries in order to establish a set of dimensions by which national culture could be defined. Hofstede (1994) has defined five dimensions of national culture:

- Power distance: the extent to which people accept that power is – and should be – distributed unequally.
- Individualism: the extent to which people see themselves as separate from others in society.

- Toughness: the extent to which achievement and success are valued. (Hofstede refers to this dimension as 'masculinity' – however, 'toughness' seems a more appropriate descriptor given the definition of the dimension – see Hofstede (1994) for a detailed description of the dimension.)
- Uncertainty avoidance: the extent to which people feel threatened by ambiguity.
- Long-term orientation: the extent to which people are future orientated.

Following a questionnaire-based survey, Hofstede scored the cultures of over fifty countries with respect to each dimension. A cluster analysis suggests that, broadly, national cultures can be divided into eight separate clusters, as follows:

Democrats: tough, short-termist cultures in which there is a very strong emphasis on individual expression. People in these cultures tend to be comfortable with uncertainty and have little respect for authority (e.g. USA, UK).

Meritocrats: tend to be uncomfortable with uncertainty and put less emphasis on individuality than do democrats. Otherwise, the values are similar to those in democratic cultures (e.g. Germany, Austria).

Egalitarians: extremely tender (i.e. the opposite of tough) cultures, with little pressure to 'get ahead'. Otherwise similar to democratic cultures (e.g. the Netherlands, Sweden).

Supportives: tender cultures that are very uncomfortable with uncertainty. These cultures have respect for authority and are not particularly individualistic (e.g. France, Spain).

Libertarians: tough, collectivist cultures, with little respect for authority. These cultures are comfortable with uncertainty and people are encouraged to succeed on their own terms (e.g. Jamaica).

Planners: very tough, very future-orientated cultures, with a strong dislike of uncertainty. Moderately collectivist with a moderate respect for authority (e.g. Japan).

Collectivists: very collectivist with much respect for authority; future orientated and uncomfortable with uncertainty; not much pressure to 'get ahead' (e.g. Brazil, Mexico).

Authoritarian: very high respect for authority; these are collectivist cultures with moderate toughness; moderately future orientated and comfortable with uncertainty (e.g. Malaysia, Indonesia).

Hofstede's cultural dimensions have been correlated against data gathered from surveys of consumer behaviour and attitudes conducted in sixteen separate countries (Mooij 1998). On the basis of the outcomes of these correlations, links were made between the five cultural dimensions and people's preferences and tastes with respect to what a product design should communicate through its aesthetics. Some of these links are summarised in the table below.

Cultural dimension	High	Low
Power distance	High status	Youthfulness
Individuality	Expressiveness	Familiarity
Toughness	Performance	Artistry
Uncertainty avoidance	Reliability	Novelty
Long-term orientation	Timelessness	Fashionableness

Mooij (1998) illustrates these links with a series of examples in the context of car design. In tough cultures, for example, people are likely to be more concerned about the performance aspects of their cars than in tender cultures. When people in a number of different countries were asked how big the engine of their car was, a negative and highly significant correlation was found between toughness and the percentage of people who did not know the size of their car's engine. In other words, people in tough cultures are far more likely to know their car's engine size than are people in tender cultures. This, then, seems to support the assertion that product performance is of a greater interest or importance to people in tough cultures than to people in tender ones. This assertion is further supported by the outcomes of a survey conducted by Eurodata (reported by Mooij 1998) which suggests that people in tender cultures were more likely than those in tough cultures to choose a car on the basis of whether they felt it looked nice, whereas, again, those in tough cultures were more likely to buy on the basis of performance.

This would suggest that, when designing for tough cultures, manufacturers should employ aesthetics that emphasise the power of the vehicle, whilst when designing for tender cultures, it may be more important to ensure that the aesthetics fit the tastes and lifestyles of the people for whom the car is designed. The links between aesthetics and culture are explored in some depth by Macdonald (1999).

Some products, through their history and contexts of use, seem to acquire particular ideological associations. The Harley-Davidson motorcycle is an example of this. These motorcycles became associated with a non-conformist, rebel image through their association with the Hell's Angels. This was not an association that was encouraged by Harley-Davidson. Indeed, in the early days of the Hell's Angels, the company refused to supply parts to shops known to service Hell's Angels' motorcycles (Scott 1995).

Nevertheless, it has been an association from which the company has, in the long run, benefited. Indeed, the image of Harley-Davidson motorcycles that developed through this association has been central to their current commercial success.

In parallel to their rebel image Harley-Davidson have also developed something of an all-American image. Again, this was something that was not necessarily deliberately developed by the company, but rather something that originally arose out of a combination of circumstances, the first of which was the USA Army's decision to use Harley-Davidsons as troop motorbikes during the Second World War. When the USA motorcycle market became flooded with Japanese imports in the 1970s and 1980s, many saw Harley-Davidson as the lone USA company, struggling against this 'foreign invasion' of their markets – again strengthening the all-American image (Scott 1995).

It is, perhaps, the all-American part of the Harley-Davidson image that helps to give the bike such a wide appeal, helping to balance the more extreme associations from the link to the Hell's Angels. This means that those riding Harleys can see themselves as rebels, whilst, at the same time, being reassured that there is nothing 'un-American' about owning such a bike.

Harley-Davidson have experienced a dramatic upturn in commercial fortunes in recent years. Indeed, the value of the company's shares has increased thirty-fold since 1986 (Hardy 1998). It seems that Harley-Davidsons are now finding favour amongst the middle-aged professional classes. People, who saw themselves as being rebellious when they were younger, may now feel frustrated by the apparent mundanity of their lives. Owning a Harley-Davidson can be an outlet for the rebellious side of their nature. After a hard week working as an accountant or a solicitor, people may enjoy getting on their Harley and speeding through the countryside. Not only can this be exhilarating in itself, but it can also reinforce a more exciting, less conformist self-image.

Another vehicle that seems to have 'accidentally' become associated with a country – in this case the UK – is the Mini. This car was launched in the UK in response to the Suez crisis of 1956; the idea being that a small car would be more efficient in terms of fuel consumption – an important financial consideration at a time when oil prices were very high (Woodham 1997).

In the event, by the time the Austin Mini came to the market in 1959, Britain was heading for the boom times of the 1960s. Austerity was the last thing on the consumer's mind as the country entered arguably the most optimistic decade in its history. On the face of it the Mini appeared doomed to failure. It certainly didn't seem to offer any benefits in terms of 'functionality' or 'usability'. Indeed, sacrifices of space and performance had been made in the interest of fuel efficiency, a quality that, by the time of its release on the market, appeared to be redundant. Despite this the Mini went on to be one of the most successful British cars of all time. Indeed, it became an icon of Britain in the 1960s. Ironically, the 'sensible' fuel-efficient economy car became a symbol of urban chic, driven by rock stars, the young and the beautiful. It was almost a 'star' in its own right. The Kings Road in London was full of Minis during the 1960s and the car was even featured in films such as *The Italian Job* (made by Paramount Pictures in 1969).

Ironically, the car was designed by a Turk, Alec Issigonis – however, the car's success was largely due to its 'Britishness' in a decade when Britain was making a huge impact on popular culture. Most of this impact came from what was known as 'Youthquake' – a loose 'movement' of creative young people such as the Beatles, the Rolling Stones and Mary Quant. These people transformed the international music and fashion scenes, and did so in a way that put a great deal of emphasis on the value of youth. This had the effect of elevating youth as something to be prized, rather in the same way that material status was in the 1950s (Steele 1997).

The Mini was strongly associated with youth and became the young person's car of the decade. This may have been in part due to its styling, which was very different from anything else on the market at the time and which had something of a 'fun' feel to it. However, the low cost of the car almost certainly played a role here too. The Mini was one of the lowest priced cars available on the market at the time, and this made it affordable for younger people of all social classes. Young working-class people may have started work at the age of sixteen and, in the booming economic climate of the time, might soon have been able to afford to buy a Mini. Meanwhile, the children of wealthy families may have been able to persuade their parents to buy the car for them. As the Mini became associated with youth, and as youth became so highly prized, older people, desiring the youthful associations, also started buying the car, eventually making it one of the best selling cars of all time.

Other products appear to develop particular lifestyle associations through their functional properties. An example of such a product is the Zippo cigarette lighter. This product is designed so that it gives a very strong and powerful flame that will not blow out even in the windiest of conditions. This functional quality makes the lighter ideal for any sort of outdoor use and has led to the association of the product with a broad series of lifestyles. One such lifestyle is that of the all-American great-outdoors man – the type of lifestyle embodied in the Marlboro cigarette advertisements. However, the Zippo lighter has also become something of a symbol of the hippie culture – associated with the camper van and music festival scene. Here the stereotype is of people sitting out on the grass, lighting joints. The Zippo, then, has built up a broad constituency of customers, based on one particularly strong functional benefit – the ability to work outdoors in windy conditions.

A trend that is currently having a major influence on design is what Popcorn (1996) refers to as 'fantasy adventure'. Fantasy adventure is associated with the desire to break out of what some may see as the mundanity of everyday life. It is reflected in the rugged design of products such as the Jeep automobile and other off-road vehicles, and the G-Shock watch. A person may drive to the office in a Jeep and own a G-Shock timepiece – a watch designed to withstand the rigours of a vigorous outdoor life, yet they may, in fact, lead a very sedentary life. Owning such products, Popcorn suggests, can provide people with a 'safe thrill', evoking associations of the great outdoors without exposing the owner to the rigours, even dangers, which might be associated with an outdoor lifestyle.

A social trend that has significant implications for product design is the changing role and attitudes of women. During the last few decades, the role of women – Western women at least – has changed dramatically. Far more women are securing highly paid professional jobs and less women are taking on the role of full-time housewives. This has brought with it a spectrum of new and different ways in which women think of themselves and in which they establish self-identities. For example, some women may identify themselves primarily by their careers, whilst others may still see themselves primarily as mothers and homemakers. This may have an effect on, for example, attitudes towards household tasks and the products associated with these. For some, doing household tasks may be seen as something central to their identity. Perhaps, for example, it is seen as a way of showing care to loved ones – perhaps seen as a major element of being a good wife and mother.

Others may see household tasks as demeaning chores – totally at odds with their identity as modern, progressive women. Far from being associated with being a good wife and mother, they may regard the expectation that women should be responsible for such tasks as a sign of an unhealthy, unbalanced relationship with their partner and as a poor example to their daughters and sons. It is important for designers to be aware that these values are likely to be associated with the norms of different cultures. In North America and Northern Europe, for example, women tend to hold far more modern concepts of femininity than do women in South America and Southern Europe. This can have important implications for the design of household products in terms of how particular design statements will be accepted.

Increasingly, manufacturers are creating household products in designs that contain an element of humour. At Domotechnica 1999, Europe's premier house-wares show, manufacturers were displaying an array of kitchen appliances, vacuum cleaners and irons in fun forms and with playful colour executions. This reflects the current tastes in Northern Europe and North America. Many women in these societies see household tasks as chores and enjoy products that, through their fun designs, can go some way towards brightening up a dull experience.

Such products may not be so appreciated in Southern Europe and

America. To the women in these cultures they may be seen as frivolous, even disrespectful. To these women, performing household tasks may be seen as an important role in their lives. They may see humorous designs as mocking or trivialising this role and may thus be inclined to prefer designs that look more rational or serious.

A product type, the customer base of which has been extended through re-styling, is the handgun. Historically, the customer base for handguns has been heavily male dominated. However, with the changing role of and attitudes of American women, gun manufacturers have increasingly seen the potential of selling handguns to women and have portrayed the issue of gun ownership as one of the last frontiers on the road to sexual equality (McKellar 1996). In particular, the range of 'LadySmith' guns, designed by Smith and Wesson, have been very commercially successful. Their marketing strategy to women is not based on the traditional macho imagery associated with guns, but rather on safety and security. Their sales brochure contains imagery of women looking after their children and seems to promote the idea that owning a gun is a way in which a woman can take responsibility for her own security and that of her children. Here is a written excerpt from the brochure:

> [A]s more women have entered the job market, become heads of households, purchased their own homes, they've taken on a whole new set of responsibilities. For their own decisions. For their own lives. For their own – and their families' – security. And security has come to mean more than a good income and a comfortable place to live. It's come to mean safety.
>
> (quoted in McKellar 1996: 74)

Here, then, guns are depicted as defenders of female equality.

The LadySmith was introduced in 1990 and is a 9mm semi-automatic. In an effort to make the gun appear more feminine than others in their range, Smith and Wesson redesigned the barrel of the gun, making it smoother and more streamlined. They also gave the LadySmith a different colour scheme. For example, the pistol is not black, like others produced by them, but rather is manufactured in a fawn/grey colour. Finally the frame of the gun is engraved with LadySmith, which is executed in an elegant flowing typeface – a marked contrast to the functional block lettering found on their other models (McKellar 1996). It seems, then, that the manufacturers have used the aesthetic properties of the gun to give it the feel of something approaching a fashion accessory – something that would not look out of place in a handbag.

A phenomenon that has potentially important implications in terms of people's tastes and values is 'postmodernism'. Postmodernists contend that universally held truths and value systems – known as metanarratives – are disappearing from society (Strinati 1995). The result of this, they suggest, is that classifications and value judgements are becoming less meaningful. In the context of popular culture this has had the effect of blurring the distinction both between popular culture and high art, and between different genres of popular culture. *Blade Runner* has been widely cited as an example of an archetypal postmodern film. This film mixes science-fiction, thriller, detective story and social observation genres in its story-line. Similarly, the postmodern person may 'eat in a luxurious restaurant one day and in a cheap pizzeria the next and may wear an expensive Rolex watch with a cheap pair of sneakers' (Marzano 1993: 7).

An implication of postmodernism is that people's tastes and values may become increasingly difficult to predict. This presents a challenge for human-factors specialists and others concerned with fitting the product to the person. Making assumptions about people's lifestyles and values on the basis of their demographic characteristics is becoming increasingly difficult. Methodologies for gathering information about these lifestyles and values, and for translating these into design decisions, will be discussed in Chapters 3 and 4.

3

CREATING PLEASURABLE PRODUCTS

In the last chapter a framework was outlined that, it was suggested, could be used for structuring thought when approaching the issue of pleasure with products. A number of examples demonstrated how products could be pleasurable, or displeasurable, by providing, or failing to provide, particular benefits to those who experienced them. Looking back at those examples, it seems probable that some of the benefits provided by these products were the result of conscious design decisions aimed at providing the particular benefit. Other products, however, may have carried special benefits as a result of 'happy accidents' – perhaps because of the historical and social context in which the product was introduced, or because the product just happened to 'click' with a particular group of people.

The aim of this chapter is to demonstrate an approach to creating pleasurable products 'by design' rather than 'by accident'. The chapter will look at how to understand people holistically and, having done so, how to understand what benefits particular people would want from particular products. The chapter will then look at how to move from a product benefits specification to the creation of a design that fulfils this specification.

Understanding people holistically

The starting point for pleasure-based approaches to human factors is to understand the people for whom the product is to be designed. Traditional, usability-based approaches to human factors tend to look at people in terms of their cognitive and physical abilities. The assumption behind usability-based approaches tends to be that if a product is designed so that the demands on the user of interacting with it fall comfortably within his or her physical and cognitive abilities, then the product will be usable for this person.

Whilst the benefits of design for usability are not in dispute, usability-based approaches tend to encourage those involved in product creation to take a rather limited view of people. Looking at people merely as 'users' may create a paradigm in which the person is seen merely as a component within a working system – a system consisting of a product, a user, a task and a context of use. It has been argued, in Chapter 1 of this book, that such approaches are 'dehumanising', ignoring, as they tend to do, the very factors that make people human – for example, their hopes, fears, dreams, aspirations, principles and tastes.

In Chapter 2, for the purposes of illustrating the four-pleasure framework, a description was given of a fictitious character, Janet Peters. In the course of describing this character the following information was given (see Figure 3.1).

Information about 'Janet Peters'

- She is a woman
- She is British
- She is a 23-year-old
- She is an accountant
- She lives in Reading (near London)
- She has a university degree
- She has no physical or cognitive disabilities
- She lives alone
- She is a 'yuppie'
- She has a boyfriend
- She finds it important to stay 'in shape'
- She enjoys physical relaxation
- She likes drinking with her friends
- Personal relationships – with her friends, family and loved ones – are important to her
- She likes to feel glamorous
- Social status is important to her
- Her work can make her feel stressed
- She hates being bored
- She enjoys feeling a sense of achievement
- She believes that it is important to behave decently towards others
- She has a sense of social responsibility
- She admires moral leaders

Figure 3.1   Information given in Chapter 2 about 'Janet Peters'

Imagine that a product development team was developing a product for someone like Janet Peters and imagine that a human-factors specialist was charged with writing a product specification which would ensure that the product would 'fit' this person. For the sake of this example, imagine that the product was a mobile phone.

If the specialist was to take a 'traditional' usability-based approach, then he or she would probably be interested in the following information, because it would relate to the physical and cognitive characteristics of the person (see Table 3.1).

Table 3.1 Information about 'Janet Peters' that may be of interest when taking a
usability-based approach to the design of a mobile phone

Information of possible interest	Cognitive/physical implications
She is a woman	Anthropometrics and physical abilities
She is British	Anthropometrics
She is a 23-year-old	Cognitive and physical abilities
She has a university degree	Cognitive abilities
She has no physical or cognitive disabilities	Cognitive and physical abilities

So, the information that may be of interest is that which pertains to her
physically and cognitively. The information that pertains to anthropometrics
may, for example, give an indication of how big the phone should be in
order that she can hold it comfortably and about what the distance between
the earpiece and the mouthpiece should be. The information pertaining to
cognitive abilities may give an indication about her ability to cope with
different interaction styles, in particular where the interaction style may
require the learning and retention of, for example, combinations of key
presses. Finally, the information pertaining to her physical abilities would
give an indication about, for example, whether she could operate the phone
using conventional keys or whether some other technology, such as speech
technology, would be required.

A pleasure-based approach to fitting the product to the person would,
however, require a far richer picture of the person for whom the product is
to be designed. Table 3.2 lists the information that may be of interest when
taking a pleasure-based approach to fitting the product to the person.

Pleasure-based approaches still include looking at usability issues, so the
cognitive and physical issues, including anthropometrics, are still important.
However, because such approaches also take into account fitting the product
to the person's lifestyle, there are many more issues that need to be considered. For example, because this person enjoys the yuppie lifestyle and finds
status to be very important, the phone may have a role for her as a status
symbol – she may want others to be able to see that it is an expensive
product and may want it to reinforce her self-image as someone who is a
successful person. Her enjoyment of looking glamorous may also mean that
she would wish the phone to have certain decorative properties. Similarly,
taste and style issues associated with being a young woman might (or might
not) mean that this person would prefer certain types of aesthetic over what
may typically be appreciated by men or by older people.

Table 3.2 Information about 'Janet Peters' that may be of interest when taking a pleasure-based approach to the design of a mobile phone

Information of possible interest	Possible implication of information
She is a woman	Anthropometrics, physical abilities, taste preferences
She is British	Anthropometrics, taste preferences
She is a 23-year-old	Cognitive and physical abilities, taste preferences
She has a university degree	Cognitive abilities, taste preferences
She has no physical or cognitive disabilities	Cognitive and physical abilities
She lives alone	Context of use
She is a 'yuppie'	Identity issues
She likes drinking with her friends	Context of use
Personal relationships – with her friends, family and loved ones – are important to her	Context of use
She likes to feel glamorous	Identity issues
Social status is important to her	Identity issues
She hates being bored	Taste preferences
She enjoys feeling a sense of achievement	Identity issues
She has a sense of social responsibility	Environmental issues

Because this person finds aspects of both her work and private life very important, it might be reasonable to expect that she would wish to use the phone in both of these contexts. This may have implications for both the aesthetics and functionality of the phone. In the work context she may see the phone as a professional tool, whilst in the context of her private life she may wish to think of it as a domestic appliance – the design team may consider how an aesthetic could be developed that would make the phone suitable for both of these contexts. Because she may use the phone as a work tool it might be appropriate to include functions such as conference calling, which may be useful in this context. Finally, her concern with environmental issues may have implications for, for example, the materials from which the phone is to be manufactured.

The speculations above are just that – speculations. The example has been given in order to illustrate the sorts of personal details that may be important and to give examples of the sorts of broad implications that these may have for the design. In Chapter 4, methods and techniques will be outlined that can help the human-factors specialist to gather accurate holistic information about the people for whom a product is being designed, information about the implications of this for specifying the benefits that a product should carry and information about the properties that a design must have in order to deliver these benefits.

The following sections of this chapter, meanwhile, will outline a structured approach to pleasure-based human factors within which such methods and techniques can be used in order to create pleasurable products.

People characteristics

The first stage in a pleasure-based approach to human factors is to gain a holistic understanding of the people for whom a product is being designed – the people who will 'experience' the product. The sections below give examples of clusters of characteristics that can differ between people. Most of these are likely to have some influence on what a person will require from at least some types of product, and examples of these are given. Once again, the four-pleasure framework has been used as a means of structuring thought with respect to people characteristics.

Once again, it is important to note that there is no claim that the clustering of the characteristics under each of these four headings represents some theoretical 'truth' about the way in which people characteristics should be classified. The assertion, then, is simply that this way of clustering characteristics can prove useful in structuring thought about people and that it helps in building a comprehensive and holistic picture of people's requirements for products.

Sometimes the clusters of characteristics may overlap and it may not always be clear to which cluster a particular characteristic belongs. This is not necessarily important. What is important is that the section headings and descriptions act as prompts to help those involved in understanding users to take a holistic approach.

It is also important to note that people characteristics that fall under one heading may influence product benefit requirements that fall under other headings. For example, a physical disability (a physio-characteristic) may have implications for people's social lives that can lead to product requirements which fit under the socio-pleasure heading. Examples of how people characteristics may affect product requirements are given below. When trying to understand people holistically, the human-factors specialist may have to move away from the 'scientific' paradigm traditionally associated with human-factors approaches. For example, not all of these characteristics may be easy to quantify – some are, but not all. Understanding people in this sense may require a degree of empathy and intuition on the part of the analyst (Macdonald 1998). Rather than seeing himself or herself as a 'scientist' studying 'subjects', it may be more appropriate for the analyst to approach people as a 'detective' – looking for clues in order to piece together an overall picture of the people for whom the product is being designed.

Physio-characteristics

These characteristics are to do with the body – for example, the senses, the musculo-skeletal system and the size and appearance of the body.

Special advantages

People may have special physical advantages, for example skills – either learned or inherent – which may give them advantages in particular situations, such as performing a particular task. For example, an advantage such as strength, quick reflexes or exceptional dexterity may be helpful in sport.

Knowing about the special advantages associated with a particular population can give the opportunity to exploit these positively in the design of products created on behalf of that population. One area in which this may be particularly relevant is in the design of military hardware. Because military hardware tends to be operated by young, fit, strong men and women, it may be reasonable to expect this user group to be able to operate the products under conditions that would not be practical for a user population that included older people or people who were not especially fit or strong. This might mean, for example, that military personnel might be able to cope with hot, cramped conditions within vehicles, or may be able to carry equipment that for others would not be portable.

Special disadvantages

Some people may have conditions that leave them, permanently or temporarily, at a physiological disadvantage. These include disadvantages associated with injuries, diseases, allergies or illnesses, as well as permanent conditions such as blindness, deafness or physical disability. There are also other temporary conditions – such as pregnancy – which may also bring physiological disadvantages in particular situations.

Clearly, the nature of a person's disability may have important consequences for design. Human factors have a tradition of contributing to design for the disabled. In particular, in terms of making products for the disabled usable. For example, Zajicek and her associates have developed a speech-based interface that enables blind people to interact with the World Wide Web (Zajicek *et al.* 1999).

Designing for people with disabilities is becoming a very important issue within manufacturing industry. The average age of the population has increased very sharply in recent years, especially in the Western world (Coleman 1999). With this increase in the age of the population comes an increase in the percentage of people who have disabilities of one sort or another. Indeed, this section of the population is now so large that many manufacturers see it as a commercial imperative that their products can be used by disabled people. This has tended to lead to 'inclusive-design' approaches – the design of products that are usable for people with and without disabilities.

Musculo-skeletal characteristics

These characteristics are those associated with the skeleton and the muscles. Physical strength is one of these characteristics. Others include flexibility/stiffness, mobility, motor control, sidedness (left, right or ambidextrous) and athleticism.

These characteristics tend to change over the course of a lifetime. A characteristic such as, for example, motor control may tend to develop until people are in their late teens or early twenties and then – in the usual course of life – deteriorate slowly from then on. The deterioration may not necessarily go on until the person becomes 'disabled', but, nevertheless, it may be important to take such factors into account when designing for the older person. For example, consider a product containing a keypad, such as a telephone. Whilst a younger person with highly developed motor control may have no problems with small, closely spaced keys, an older person with less motor control may have considerable difficulties with such a layout. Larger and more widely spaced keys might be required for the older user.

Similarly, children may have difficulty with tasks that require a high level of motor control. For example, many children have trouble using computer input devices. For example, moving a cursor over an on-screen icon using a mouse can demand quite fine control. When designing computer input devices for children, it tends to be advisable to 'gear' these far less responsively than would be the case with input devices for use by adults. Gearing refers to the ratio of the distance that the user moves the mouse to the distance that the cursor moves on the screen. If the gearing is 'sensitive' this means that a comparatively small movement of the mouse will result in a comparatively large movement of the cursor – something that children tend to struggle with.

External body characteristics

These refer to the 'external' characteristics of the body. These include height, weight and body shape, anthropometric dimensions and facial features. Hair, eye and skin colour are also included here. The condition of visible characteristics, for example skin condition and the condition of a person's teeth, would also be included in this cluster.

External body characteristics can vary fairly systematically across populations. For example, men tend to be, on average, taller and heavier than women. Similarly, certain ethnic groups may differ anthropometrically from others. This sort of information is often essential when designing products for sale to different populations. It seems to be an accepted norm within human factors that products should be designed so that they fit – in terms of, for example, strength and anthropometrics – 95 per cent of the people for whom they are designed (see, for example, Grandjean 1988). Markets for products are usually defined in terms of such general characteristics as those in this cluster. For example, the market for a car may be defined in terms such as 'men and women in Europe and Asia'. Knowing this it is possible to deduce, for example, the critical dimensions for ensuring that 95 per cent of this population can reach the pedals and at the same time have a good view of the road. This can be done using anthropometric tables of the sort found in *Bodyspace* (Pheasant 1986).

Body personalisation

People can make alterations to their bodies in order to 'personalise' them in some way. This might include, for example, decisions about hairstyles, bodily and facial hair, piercing, tattooing and jewellery. Decisions about wearing glasses or contact lenses would also be amongst the factors considered in this cluster. Plastic surgery offers the possibility of more extreme forms of body personalisation and would also be a factor that might be considered within this cluster.

Although a physiological characteristic, body personalisation is, perhaps, most often a means of social and ideological expression. The human-factors specialist, then, might consider that the way in which a person has personalised his or her body offers clues about the wider values which that person holds.

Some products are designed specifically for body personalisation – for example, shavers, beard trimmers, hair-care products, sun beds and other beauty products. Perhaps, then, the design of products created for body personalisation should not only be supportive of the task for which they were created, but should also express something of the values of the person for whom the product is being designed.

Consider, for example, a female depilation product, such as an electric razor for use on the legs and under the arms. It may be, for example, that women who are particularly concerned with this aspect of their appearance – concerned enough to purchase and use an electric razor – particularly enjoy expressing their femininity. Perhaps, then, this should be reflected in the design of the product – for example, in the form and colouring of the product.

Physical environment

People will be affected by their physical environments. This includes factors such as the temperature and humidity of the environment. The clothes that a person is wearing may also be considered as – in effect – part of the environment that is surrounding him or her as these may affect both temperature and the ability of the body to respond to temperature. Some clothes – particularly, perhaps, protective clothing – may also constrain the degree of freedom of movement that a person has. Similarly, in the case of headgear, a person's range of vision may be restricted.

Another environmental issue that can, in some circumstances, prove significant is that of environmental noise. In particular, there are many noisy workplaces. When designing, for example, alarm systems for such workplaces, it may be sensible to back up any audible warning signal with a visual signal. This might mean, for example, that in addition to, say, a fire alarm siren, a flashing red light could be used to warn those in the workplace of any impending danger.

Physical dependencies

Some people have addictive dependencies that affect the way in which they live their lives. These may include dependencies on, for example, tobacco, drugs (legal or illegal) or alcohol. Eating disorders, such as compulsive eating or bulimia, would also come into this category.

A consensus seems to be emerging amongst the medical profession that not only smoking, but also breathing the smoke of others, can be detrimental to health. This has led, for example, to a ban on smoking in many public environments – particularly in the USA. Even smoking in bars is banned in some parts of California. Clearly, many smokers may be concerned about the effects that their smoking may have on their families – particularly their children. A business opportunity associated with this is the development of air cleaners so that they can filter dangerous carcinogens from the atmosphere. Originally air cleaners were designed to get rid of dust and smells; the addition of carcinogen extractors represents an example of adding a requirement to the product based on an understanding of an emerging need.

The styling of such products can be very important in determining how well they are accepted by those for whom they are designed. For example, some smokers may not like the idea of having a medical looking product in their homes – this might seem to suggest that their tobacco addiction is a terrible medical condition from which others need to be protected. Rather, it may be more appropriate to style the product as a piece of household furniture. The message here might be that smoking is a lifestyle choice that, with an air cleaner in the house, need not have any detrimental effect on others.

Reaction to the physical environment

Different people may react in different ways to the same physical environment. For example, some people seem to be able to adapt to heat or cold better than others. The same may be true for environments containing, for example, smoke, dust or chemicals.

This can be an important issue when considering the design of products associated with household cleaning – indeed it has been picked up on by a number of vacuum cleaner manufacturers. Vacuum cleaners are now offered with a range of special features and functions that enable them to be highly efficient in removing a variety of potential allergens from the atmosphere, including pet hair, dust mites, house dust and pollen.

Socio-characteristics

These are to do with – in the widest sense – a person's relationships with others. They include, for example, relationships with friends and loved ones, with colleagues and peers, and with society as a whole

Sociological characteristics

These include the country and culture that a person lives in and the values and customs associated with these. Political considerations, such as human rights, may play a role here. Characteristics of the neighbourhood and home in which a person lives may also be included in this cluster.

This issue can be particularly significant in the context of products designed for women, particularly those associated with personal care. In the Western world such products are often designed in order to emphasise femininity and sensuality. The Ladyshave, as pictured in Figure 3.2, has an overtly feminine form and friendly looking graphics, indicating which function is for the legs and which for the underarms. Such a use of form and graphics may be considered inappropriate in, for example, the Muslim-governed countries of the Middle East. Here, female 'modesty' may be strongly encouraged, even imposed by law. For example, some women's magazines, such as *Cosmopolitan*, are banned in a number of Muslim countries because they are considered to be pornographic. Products whose forms appear to celebrate feminine sensuality may be inappropriate in such countries. An aesthetic that gave the product the appearance of a 'sub-medical' tool might be more appropriate.

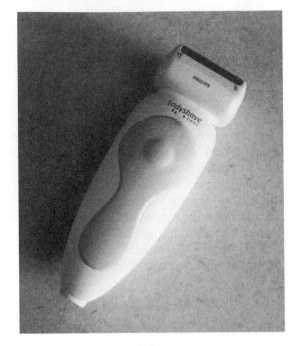

Figure 3.2   Philips Ladyshave

Status

This is to do with a person's 'standing' in society – how he or she is perceived by others. Socio-economic status may play a role here – people in the 'top' professions may gain respect from others. However, cultural status may also be important: for example, being a 'cool' person to be seen with, driving a flashy car or being knowledgeable about the arts. Sometimes status may be gained from being part of a social group. This could range from being a member of a professional society to being a member of a street gang. Titles can also be a source of, or a reflection of, social status – for example: Lord, Sir, Baroness, Lady, Doctor or Professor.

Different people have different attitudes towards status. For example, some high-status people may be very proud of their status and want to show it off. For others status may not be an issue at all and others may want to play their status down. This is a very important issue in the context of the design of a huge variety of products. Almost any product that is seen by others has the potential to be a status symbol – from a kettle to a car. Typically, materials, finishings and product decoration are used as a means of adding status to a product. For example, a return to the use of 'noble' materials such as metals and woods, as opposed to plastics, can add status to a product.

As a simple example, consider the design of pens. Pens are a product whose design requirements may well be influenced by the need to carry status. Clearly, a gold-plated pen, or even a pen fabricated from aluminium or steel, is likely to create far more of an impression of status than an equivalent pen fabricated from plastic.

It is important to note that status is not only associated with material success. Another type of status is connected with culture – appreciating the 'right' things. This is what Bourdieu (1979) refers to as cultural status. Returning to the example of the pen, imagine that a pen was designed with gold plating and a diamond-encrusted clip – in other words, imagine that this pen was designed to be extremely ostentatious. Whilst owning this pen would certainly be a symbol of material status, it is questionable as to whether it is a symbol of cultural status. Some people, for example, might consider such extreme ostentatiousness as being 'vulgar'.

Social self-image

Aside from how others see them, people have their own ideas about their sense of social identity and status. Some may be self-confident with a high level of self-esteem. However, if this becomes exaggerated it may lead to the development of an anti-social trait such as arrogance or snobbishness. On the other hand, those that lack self-esteem may be filled with self-doubt. Some people may continually feel the need to impress in order to maintain or build a positive self-image.

Many products can have a role as 'social accessories' – helping to generate or maintain a particular image. People may want to own or use products whose design tallies with their image of themselves – how they think others see them, or how they would wish others to see them. Consider, for example, a product such as a watch. Watches come in a great variety of designs, projecting a variety of different qualities that, in turn, may project different images of the wearer. A 'gadget' watch, such as a digital watch with many functions or an analogue watch with a variety of dials, may give an impression of the wearer as someone who is happy dealing with complex technology, whilst a simple, one-dial analogue watch may give an impression of someone who appreciates purity and simplicity. Other watches may be designed so as to be chunky and 'heavy duty', perhaps giving the impression that the wearer leads a tough, adventurous lifestyle.

Social relations

Social relationships are another aspect that may play a very important part in defining a person's social characteristics. These may include friends, family and loved ones. Social preferences may also play a role here. For example, some people may tend to prefer the company of men, whilst others may prefer to be with women; some may prefer to mix with people of their own age group, profession or status, whilst others may prefer more mixed company. Some social relationships may be 'formalised' through belonging to social organisations, for example the Women's Institute or a gentlemen's club.

A significant issue here is the living circumstances of a person. Do they live in a large family or do they live alone? This can have a number of practical implications for the design of products, particularly those that are used in the household. Perhaps the most obvious connotations are those associated with the capacity of products, for example food and beverage preparation products, such as coffee-makers, food processors and blenders.

However, there are also emotional design issues connected with social relations. For example, if a product will be used and seen by a number of different people – for example mum, dad and the children – it may be more appropriate to use somewhat more neutral aesthetics than if the product is for one person only. Where a product is for one person, he or she may regard the choice of product as an opportunity to express his or her personal taste – this may not be such an issue in a product designed for communal use.

Social labels

Sometimes people may have social 'labels'. These are the characteristics of people that others may use to make assumptions about their social attributes. One potential source of social labelling is the person's visible characteristics. This might include, for example, gender, age and ethnic origin. Other characteristics that can lead to labelling include, for example, a person's accent, name or nationality. People who are labelled in this way may sometimes be proud of the way in which others see them, but may equally feel that they are being stereotyped, patronised or discriminated against depending on the attitudes of the 'labellers'. Sometimes people may consciously or unconsciously label themselves – for example, through clothing, hairstyle or outward signs of affiliation with a particular group.

Again, different people may have different attitudes towards social labelling. Some may enjoy being seen as part of a particular social group and may wish to exhibit the trappings associated with that group. Others, however, may regard social labels as annoying or offensive and may deliberately steer clear of any product that they feel is likely to lead to them becoming labelled in a particular way.

The flashy aesthetics associated with 'yuppie' products are an issue here. Some people may enjoy being seen as a member of this group and may be attracted to products that have such aesthetics. On the other hand, some, of similar age and professional and financial status, may consciously avoid them.

An issue associated with many of the 'new elderly' is a resistance to be labelled as elderly. This is a particularly Western phenomenon, stemming from the fact that many of today's 'sixty-somethings' were in their twenties and thirties during the 1960s – a time when youth was very highly valued (Steele 1997). This can lead to a great reluctance amongst this generation to think of themselves as 'old' or 'elderly'. People who have these attitudes are unlikely to be attracted by the sorts of 'sensible' aesthetics traditionally associated with appealing to the older generation. They may enjoy exciting, youthful, vibrant aesthetics and still be very fashion conscious. This phenomenon is noticeable in the motor industry, particularly in the high sales of sports cars to this age group – many may have dreamed of owning exciting products like this when they were younger. Now after a lifetime of working and with the children having left home, they can afford them and can 'grow old disgracefully'.

Social personality traits

Many personality traits may have an effect on the way that people relate to others socially. For example, some people may be particularly generous or have a strong sense of community and social responsibility. Others may be particularly caring or loving. A trait that can have a significant effect on social behaviour is that of conformity versus rebelliousness.

Again, this can have a significant influence on the sorts of aesthetics that people are likely to appreciate in products. In particular, the phenomenon of 'suppressed rebelliousness' is something that has had a significant effect on the market-place in recent years. Many people may have had rebellious ideas in their youth, but now feel that they are living very conformist lifestyles: working nine-to-five in the professions; going to work in a suit and tie every day; living in suburbia. In their free time people may want to react against this and express a degree of 'safe rebellion'. Harley-Davidson attribute much of their recent revival to this phenomenon. These days there are probably as many accountants as there are Hell's Angels riding Harleys. Because the bike has, over the years – perhaps because of its associations with the Hell's Angels – become an icon of rebellion, people may feel that just riding one gives a feeling of rebelling (Hardy 1998). This phenomenon has been noted by many Japanese manufacturers who now offer a range of bikes that carry the Harley-Davidson aesthetic.

Social lifestyles

Different people may have different social lifestyles. For example, some may be very socially active, often going out and meeting others, whereas others may prefer to spend their time at home – perhaps reading or watching television. Some people may be family men or women, preferring to spend as much time as possible with their partners and children. Others may enjoy living 'glamorous' or 'fun-seeking' lifestyles – on the look-out for new and exciting social experiences.

71

Social lifestyles can, again, have significant influence on the sorts of aesthetics that people may appreciate in products. For example, those who aspire to cosy, family lifestyles may tend to prefer 'safe' aesthetics. The feeling behind this may be that these people may want to spend their money 'sensibly' – feeling secure in the knowledge that they have done their best for their families and that they have not been frivolous with their money. Of course, what is 'safe' is a concept that changes over time. Consider, for example, the use of colour in household products. About five years ago – in the mid-1990s – the use of strong colours in household products, from sofas to crockery, was considered somewhat avant-garde – something that would appeal to those looking for innovative, 'fun' products. Since then, however, the use of strong colour in these products has become the norm. Now the more innovative designs tend to use colour in a more subtle way.

By their very nature and by the nature of the way that perceptions of aesthetics change over time, 'fun seekers' tend to prefer more avant-garde designs. Having something 'different' can be very appealing to these people, whereas to others having something different may be seen as risky – they may wish an aesthetic to become proven and familiar before it would appeal to them.

Psycho-characteristics

These refer to the cognitive and emotional characteristics of people.

Special talents and difficulties

This cluster includes characteristics such as people's intelligence, skill and creativity. It also covers people's ability to learn and memorise, as well as their perceptual abilities. Others may suffer from mental disabilities that may result in, for example, learning difficulties or difficulties in memorising things.

This has been shown to have an implication for, for example, the structuring of menus in software packages. As people get older, short-term memory capacity tends to deteriorate (Freudenthal 1997). This may mean that older people will have greater difficulties in remembering which commands fall under which menu heading. This suggests that older users may prefer 'broad' as opposed to 'deep' menu systems. A broad menu system is one that includes many menus, each containing a comparatively small number of commands. A deep menu structure, meanwhile, is one that contains few menus, each of which contains a comparatively large number of commands. Deep menu structures have the advantage of making an interface less cluttered. There are less headings to choose from, so it is easier to find the menu that needs to be opened.

However, a potential problem with deep menu structures is that some people may have difficulty remembering which menu the required command is on. If he or she is likely to have trouble remembering this – and older people may be more inclined to have difficulties than younger people – then it may be more sensible to use a broad menu structure, despite the possible implications for clutter.

Psychological arousal

This category refers to temporary states of psychological arousal that may affect people at particular times. For example, a person may be feeling alert, stressed, tired, bored, etc.

These factors may be important in the context of products that may be used in, for example, circumstances of extremely low or high psychological arousal. Consider, for example, radar operators. These people may have to monitor radar screens for long periods of time, during which they may become bored, tired and under-stimulated. In order to design radar systems that are readable under such circumstances, it is important to have an understanding of how people's signal detection performance is likely to be affected by such factors. This is an area that has received a great deal of attention from psychologists and human-factors specialists – see, for example, Shackel (1999).

At the other extreme, consider a person trying to operate a product under circumstances of extremely high psychological arousal, perhaps brought on by fear or panic. In such a context it is important to be aware of the 'population stereotypes' that may be held by the people using the product. A population stereotype is an expectation held across a particular population about the way in which a product will work. In the USA and mainland Europe, for example, switches are typically flicked upwards in order to turn something on. In Britain, conversely, it is typical to flick a switch down to turn something on. Similarly, emergency exit signs in the USA have red lettering, whilst in Europe the lettering is green.

Population stereotypes should be taken into account when designing for particular markets. This is particularly so when products have a safety-critical aspect to them, because in the heat of an emergency people may revert to 'instinctive' behaviours. So, for example, if designing potentially hazardous industrial machinery for sale to Britain, the switching on the machine should require the operators to switch it upwards in the event of an emergency shutdown being necessary. However, if the machine was being sold to companies in continental Europe or the USA, then the switching should be down for off.

Sometimes a person may be particularly alert or sharp, whilst at other times he or she may be muddled or absent-minded. Clarity of mind might be affected by, for example, tiredness, time of day, distractions in the environment or distractions that result from doing other tasks. It may also be positively or negatively affected by drugs such as alcohol, nicotine or caffeine. A person's mood may also play a role here. For example, a person may be happy or sad, calm or anxious, aggressive or passive, delighted or outraged, etc.

A spoof article in *Viz* magazine claimed that the invention of microwave chips was putting the British fire brigade under threat. Apparently, after returning drunk from an evening in the pub people were cooking microwave chips rather than trying to cook chips in deep-fat fryers – the subsequent reduction in chip pan fires meant that there was less work for fire-fighters and job cuts might result!

Behind this spoof there lies a serious issue. It cannot be assumed that people will always be alert and sharp-minded when using products. This issue is particularly important to consider when creating safety-critical products. This may be of special importance in the context of household products. Consider, for example, a potentially safety-critical product such as an iron. Whilst a fully alert person may remember to turn the iron off after use someone who is distracted, stressed, rushed or tired may forget this. Many manufacturers have included automatic shut-offs in their irons in order to compensate for this. These shut-offs switch off the iron if it hasn't been moved for a particular period of time, or if it is left with the ironing plate down.

Other household products have been designed with features that keep the user from placing himself or herself in danger through putting their hand into the product whilst it is in operation. Examples of such features are the safety locks on the doors of washing machines and spin dryers. These locks will prevent the user from opening the door of the machines whilst they are still in operation. Similarly, most microwave ovens will automatically shut-off whilst the door is open.

Another excellent example of safe design for the careless user is the garden strimmer. The cutting tool on these is usually made from nylon thread. The properties of this material mean that it will effectively cut through grass and weeds, but that it will not cut through human skin.

The context in which some products are used can make them potential distracters from other – possibly more important – tasks. A particularly pertinent example of such a product is the car stereo. Operation of this product is a potential distracter from a far more important and safety-critical task – that of driving the car. This might mean that the demands on the design of a car stereo in terms of usability will be considerably more exacting than the demands on designing a stereo for use in the living room. In the case of a household stereo, it might be accepted that provided people could learn how to use the main functions within a reasonable period of time, then the stereo was usable. In the case of a car stereo, however, the issue would not simply be about whether people could use the stereo, but also about the level of distraction that its use had on driving.

Personality traits

This cluster is concerned with comparatively steady psychological attitudes. They are not moods that come and go comparatively quickly, but rather are part of a person's psychological make-up. Examples include being extrovert, introvert, aggressive, passive, perceptive, intuitive, etc. (Kline 1993).

Research has shown that people tend to imagine products as having personalities and that they tend to express a preference for products that they perceive as reflecting their own personalities (Jordan 1997). For example, in the context of domestic appliances, people tend to assign extrovert personalities to products that have novel forms, bright colours and shiny finishings. Meanwhile, products with geometrically strict form language tend to be seen as introvert. If people see themselves as extrovert they tend to prefer the former and if they see themselves as introvert they tend to prefer the latter. This phenomenon will be discussed in more detail in Chapter 4, where Product Personality Assignment is reported as a case study.

Self-confidence

This refers to the level of a person's self-belief. Self-confident people tend to have a positive image of themselves and a belief that they can achieve. On the other hand, people who lack self-confidence may feel less self-assured.

As well as possibly being a general trait, this trait may exist in people in relation to particular products or product types. A particularly pertinent example of this is in relation to people's attitudes towards computers. For many, computers may be 'frightening' high-tech products that, by their very nature, inspire a lack of self-confidence. Others, however, may feel very self-confident when faced with computer technology. They may enjoy mastering high-tech products and feel very comfortable doing so.

An example of a product that owes much of its success to its appeal to the former group of people is the Apple Macintosh computer. This product, with its easy-to-use direct manipulation operating system and its friendly looking icons, has, for many, 'demystified' computers. This has enabled people who may previously have been lacking self-confidence when dealing with computer technology to approach this technology in a more confident frame of mind.

Another information technology-based product in the context of which lack of self-confidence may be a barrier to usage is the ATM – the Automatic Teller Machine. Nowadays virtually all banks offer ATM services to their customers. However, many people are still reluctant to use them for fear that they will make errors that will result in embarrassment in front of others, or in accidental withdrawals from their accounts. Again, ease of use is one key to solving this problem. Now, the majority of ATMs offer the user a series of menus from which to make their choices, minimising, or eliminating, the requirement to type in command strings. In particular, where once users were expected to type in the amount of money that they wished to withdraw, they are now offered a choice from a selection of fixed amounts.

Similarly, the operation protocol for using such machines has also been altered in order to minimise the number of potential errors that a user can make. In particular, users are now asked to remove their bank cards from the machine before their money is vended. Previously, the protocol had been for the money to be vended before users were asked to remove their cards. This often led to users taking their money and walking off without their cards, which might then either be taken by someone else or retained by the machine. The new protocol seems to work better because people's reason for going to the machine in the first place is usually to withdraw money – they are, then, unlikely to forget to do this and will tend to stay at the machine until they have their money. However, once people have their money, they may feel that the transaction is complete and walk away from the machine.

Clearly, it is likely that the more successful interactions a person has with an ATM machine, the higher his or her confidence will become with respect to this technology. However, as with the Apple Macintosh, confidence may be built not only from the usability of the product, but also via the aesthetics of the interface. For example, Kim and Moon (1998) found that use of colour and graphical layout of the screen had a significant effect on how confident people felt about using automated banking systems.

Learned skills and knowledge

People's psychological characteristics also include skills that they have learned and knowledge that they have picked up. These might be general skills, such as learning a language, or specific skills, such as learning to use a particular product. Knowledge refers to facts that a person may have learned through being formally instructed or educated, or which he or she may have picked up in the course of life generally.

Where relevant, it can be important that the design of a product makes use of the skills of those who will use it. For example, if designing a product for professional usage, such as a drawing package for use by industrial designers, the functionality included and the interaction style should be decided in the light of the skills that industrial designers have and the ways in which they have been trained to work. This might mean, for example, that the way in which images are created on screen would take advantage of the users' free-hand drawing skills. This might mean allowing the user to create an image through movement of the cursor using the mouse and then 'tidying up' this image using keyed inputs to specify the sketched dimensions more precisely. On the other hand, a drawing package aimed at, for example, mechanical engineers – typically more mathematically trained, less artistic people – might rely far more heavily on the keyed input of precise dimensions right from the start of creating an image.

Sometimes relying on people's learned skills might mean that, for the sake of compatibility with other products that the person has used, it may be better to settle on a design solution that, when considered context-free, might seem sub-optimal. Imagine, for example, that a study were to show that the conventional layout of the control pedals in vehicles – clutch on the left, brake in the middle, accelerator on the right – were sub-optimal. It would still, almost certainly, be extremely unwise for a manufacturer to switch the position of these pedals on a new model!

A similar issue, and one that has received considerable attention from human-factors specialists over the years, is the layout of alpha-numeric keyboards. Human-factors specialists have long been baffled as to why the Standard American Keyboard (otherwise known as the Scholes or QWERTY Keyboard) is arranged such as it is. Experiments have shown that this arrangement is by no means optimal in terms of typing speed. Other arrangements, such as the American Simplified Keyboard (also sometimes called the Dvorak Keyboard, after its inventor), have been shown to be far more effective. Even keyboards simply arranged with the keys in alphabetical order seem to out-perform the QWERTY keyboard. Indeed, in experiments reported by Norman (1988), the performance of novice users with the QWERTY was no better than with keyboards where the keys had been arranged at random.

A number of different explanations have been offered as to why the QWERTY keyboard is laid out as it is. Norman (1988) suggests that the reason is because the levers on early typewriters were prone to jamming. This, he suggests, was more likely to happen when the typist was typing quickly and when keys in close proximity were pressed one after the other. The solution, then, was to change the location of the keys so that letters that were often typed in quick succession – such as i and e – were placed on opposite sides of the keyboard. This, then, would reduce the regularity with which jamming occurred. Others, however – for example Noyes (1998) – are not convinced by this explanation and conclude that the true explanation may never be known.

So, given that the layout of the QWERTY keyboard is sub-optimal, why does this layout persist on the vast majority of typewriter and computer keyboards, even to this day? Again, the answer is that it draws on skills that people have learned. Because so many people have already learned to type using a QWERTY keyboard, they would almost certainly suffer a fall off in performance should they try to type with a keyboard with a different layout. And, because nearly all keyboards in circulation have QWERTY keyboards, new typists will almost certainly learn to type on a keyboard with a QWERTY layout. It is, then, a seemingly unbreakable cycle. QWERTY keyboards dominate because people have learned to type on QWERTY keyboards and people learn to type on QWERTY keyboards because QWERTY keyboards dominate. This might be seen as an unfortunate cycle as it perpetuates a non-optimal solution. Nevertheless, it would take a brave – perhaps foolhardy – manufacturer to try to break it.

Ideo-characteristics

These characteristics are to do with people's values. They pertain to their tastes, morals and aspirations.

Personal ideologies

These are ideologies that a person uses – or, at least, tries to use – as a basis for personal lifestyle choices and for setting goals and aspirations. Adherence to traditional family values would be an example of the sorts of issues that come into this cluster, as would, for example, a work ethic or achievement orientation. Some personal ideologies might be more hedonistic in nature – such as decadence, epicurism or materialism. Other ideologies might emphasise 'purity' – for example, temperance or minimalism. Some may have personal ideologies that include concepts such as self-actualisation.

Personal ideologies may have an influence on the sorts of aesthetics that people appreciate in products. For example, if a person were to have a personal ideology that emphasised minimalism and purity, then he or she may be inclined to purist or minimalist design aesthetic. This, for example, was the ideology that underpinned the philosophy behind the modernist design movement, which gained momentum during the 1920s and 1930s (Hauffe 1998).

This movement may have been a reaction to the somewhat 'fussy' design aesthetics that were popular during the Victorian era, when products tended to have a lot of non-functional decoration and detailing. Indeed, the modernist movement claimed that, because their approaches spared non-functional embellishment, they were intrinsically 'honest' designs – supporting the product's function, rather than having any pretensions about the product as artwork or sculpture. Modernist designs, because they were created in order to best emphasise and support the product's function, were, inherently, regarded by their advocates as being 'timeless' – less prone to the whims of fashion than decorative designs. Today, the modernist approach is, perhaps, most famously championed by German manufacturers Braun, whose designs tend to embody a 'form follows function' philosophy.

In sharp contrast to this was the 'jet-age' aesthetic that became popular during the 1950s, particularly in the USA. This aesthetic was characterised by 'streamlining' products into aerodynamic forms – forms that, through their association with aerospace technology, helped to give products a high-tech image. This was characterised in the automobile industry by the 1950s USA 'gas-guzzlers' – huge cars, with streamlined wings and acres of chromium. Whilst there may appear to be some logic behind the application of the streamlined aesthetic to automobiles, this aesthetic was also applied to stationary household appliances, such as fridges, furniture and even clocks.

This aesthetic put emphasis on technological progress and on the need to have the latest most 'up-to-date' products. Many designers felt that the aesthetic was inherently 'dishonest'. After all, they argued, products were simply being restyled in order to look more up-to-date, whilst, in reality, their functionality and technology remained the same (Votolato 1998). Further, this type of styling was adding no advantage in terms of performance – what is the functional rationale behind streamlining a fridge? Nevertheless, such products were eagerly purchased by a newly prosperous American public, determined to have the latest and best things after the hardship and frugality of the Depression and the war years. Many enjoyed having the latest products and were keen to buy the newest things. Manufacturers capitalised on this by bringing out new designs of the same product year after year. This is the concept of built-in aesthetic obsolescence. People who enjoyed this form of consumerism would want to replace their products with the most up-to-date styles, even if their current products were still working perfectly well.

Religious beliefs

A person's religious beliefs – or lack of beliefs – are also a part of their ideology. People may belong to one of the major organised religions, such as Christianity, Islam or Judaism – or they may adhere to a moral code, such as humanism, or mystic traditions such as spiritualism. Some people may have personal religions or moral codes that they have developed for themselves. Others may be agnostics – people who are unsure about their religious beliefs – or atheists – people who are convinced that there is no God or spirit world.

To the Westerner this might seem something of an irrelevance in terms of product design. After all, it seems rather unlikely that a manufacturer would load their products with religious symbolism. In the West, religions tend to have symbols associated with them – for example, the cross is associated with Christianity, whilst the Star of David is associated with Judaism. Certainly, it would be an insensitive designer who would use such symbols to, for example, label buttons or switches on a product.

However, this issue can trip Western manufacturers up when designing products for sale to Eastern markets. The reason is that a number of Eastern religions attach a great deal of symbolic significance to colour. Whilst, for example, to the Western designer, black might seem an elegant, sophisticated colour, this may have very negative connotations within the context of some religions. On the other hand, there are other colours, for example green, which can be considered very positive within certain religious contexts.

Other religions may have particular doctrines that affect what people can eat or drink. This can have significance for the design of some products associated with the preparation or storage of food and beverages. Consider, for example, the design of a product such as a fridge. If designing for the USA market it may be sensible to design part of the in-door shelf space to hold the standard twelve-fluid-ounce American beer bottle. However, such a feature would probably meet with little enthusiasm in a Muslim country, such as Saudi Arabia, where drinking alcohol is prohibited.

Social ideology

This category refers to people's beliefs about the way in which society should conduct itself and the way in which they should interact with others in society. For example, some might regard alcohol as a 'social evil' and feel that people should be temperate or teetotal. Other social ideologies might include, for example, respect for authority, environmentalism, political correctness, care for others, technophilia or technophobia, as well as particular moral views about how society should behave.

Environmentalism is an issue that has had a major influence on product design over the last two decades. In particular, this has affected the way in which materials are used in products. People who are concerned with environmental issues may be inclined to avoid products that use environmentally harmful materials, such as plastics and PVCs. They may prefer products that use environmentally safer materials such as metals, woods, glass and the newly developed biodegradable plastics.

Other products may be environmentally unfriendly, not because of the materials used in their manufacture, but because of resources used in their running. In the automobile industry, this has been reflected in the success of fuel-economic vehicles at the expense of the 'gas-guzzler'. In the household, a product whose design has been affected by this issue has been the vacuum cleaner. The filters used in these machines may sometimes contain harmful chemicals and the need to replace the bag is a potential source of environmental damage. Over the last five years or so a number of manufacturers have tackled these issues and brought 'environmentally friendly' vacuum cleaners to the market. These may contain filters manufactured without the use of harmful materials and may not require a dustbag. Dyson vacuum cleaners are examples of designs with such characteristics.

Aesthetic values

This is to do with what people will find aesthetically pleasing – for example, a person's judgement about what is beautiful or attractive. In the context of the visual arts, for example, some people may prefer classical works and others modern works. Attitudes towards style would also be included here. For example, a person may enjoy following the latest fashions, may have a timeless sense of style, or may not be interested in style at all.

In much of the preceding discussion on people characteristics examples have been given suggesting that certain characteristics may be associated with particular aesthetic tastes. Equally, however, it may sometimes be possible to come at the issue from the other side and to characterise people *in terms of* their aesthetic tastes. Manufacturers often use this as the basis for building up 'ranges' of products. Ranges are sets of different products that have some common design element to them. For example, a range of household products might include, say, a refrigerator, a cooker, a dishwasher and a microwave oven. The range may be co-ordinated through a combination of, for example, colour, form language, materials and graphic elements – indeed, through all or any of the different aesthetic elements.

Some manufacturers may have a consistent aesthetic style across all of their products. For example, German manufacturers Braun have tended to consistently employ a modernist, functionalist aesthetic to their products. This aesthetic approach was based on the philosophies of the Bauhaus and the Ulm Academy for Design and was implemented within Braun under the guidance of Dieter Rams who started working for the company in 1955. The approach tends to be associated with a modernist, minimalist design ethic – no unnecessary embellishment, but rather forms that directly support the function of the design. A consistent approach such as this can be beneficial in helping to establish a strong brand image for a company, and in building customer loyalty amongst those who appreciate such an approach.

However, there are also associated risks with such an approach. By designing to a consistent aesthetic identity a manufacturer runs the risk that they limit their potential market to those who happen to appreciate this particular type of aesthetics.

Aspirations

Different people will have different aspirations and will wish to see themselves in different ways. Some people, for example, may wish to see themselves as successful in their career; others may regard being a good father or mother as important; others may value intellectual ability or sporting prowess. Some people like to see themselves as modern men or women, whilst others may put great emphasis on tradition. A person may be a realist, a purist, a pragmatist, a cynic, etc. People's perceptions of, for example, events, people or entities, may be 'coloured' in a particular way, depending on their own particular outlook and aspirations.

This cluster of characteristics may have considerable overlap with the characteristics within the 'social self-image' cluster. The essential difference, though, is that the characteristics in this cluster are about how people wish to see themselves, not necessarily about how they would wish others to see them. This difference may be important in terms of its effect on the range of products and contexts in which these characteristics are important.

Consider, for example, a kitchen product such as a kettle. Functionally, this is a fairly straightforward product – it boils water and that's about it. However, to some, even a simple product such as this might be seen, explicitly or implicitly, as carrying a number of values that can reflect the values of its owner. Again, with a product of this nature, the values carried by the product are likely to be implicit in the product's aesthetics. The product might, for example, look like a typical, functional kitchen tool. On the other hand, it might be designed to be an *objet d'art* – perhaps through an unusual or sophisticated form, or through the use of noble materials, such as metals rather than plastics.

A person's aspirations may have a significant influence on which aesthetics he or she prefers. For example, if a person aspired to think of himself or herself as someone with a sophisticated appreciation of design, then he or she may prefer the more interestingly styled design. On the other hand, if he or she aspired to think of him- or herself primarily as a homemaker, then more 'normal' or 'safe' design might be preferred.

Product benefits specification

In the previous section, examples were given, demonstrating how a person's characteristics might affect the benefits that he or she would require from a product in order to find that product pleasurable. In this section, the focus is on how to create a 'product benefits specification' – a list of benefits that a product should deliver to those for whom it is designed. This specification can form the basis for the proposal and evaluation of design solutions.

To illustrate what is meant by a product benefits specification, consider the following case study – originally reported by Jordan and Macdonald (1998) – which considered what the product benefits specification for a photo-camera might be. Note that the product benefits specified are derived from information or assumptions about the characteristics of the target group.

Case study part 1: creating a product benefits specification for a photo-camera

Target Group

Imagine that the target user group for the camera is Western women aged between twenty-five and thirty-five of high socio-economic status. Given that this is the target group, what are the implications in terms of product requirements? What follows is a four-pleasure analysis, suggesting some issues that may be of importance in this context and considering what the associated design requirements might be.

Physio-pleasure

A camera is a hand-held device. Clearly, then, the feeling of the camera in the hand may be a source of physio-pleasure to the user.

Product benefit: camera should feel good in the hand.

A camera is a product that people are likely to want to carry around with them. This is likely to hold for this target group as much as for any other.

Product benefit: camera should be easy to carry around.

When taking a photograph, the user will hold the camera to her face and look through the viewfinder. Here the camera may come into contact with parts of the face, in particular the side of the nose and the eyebrow. Again, the camera has come into contact with the body and should fit well against the face and be comfortable.

Product benefit: fits well and comfortably against the face.

Finally, for this example, the designers should be aware that many young women may have long fingernails. Clearly, they will not want these broken when using the camera.

Product benefit: camera should be operable without causing damage to the users' fingernails.

Socio-pleasure

A pocket camera is something that will often be used in a social context. The user may want to take photos of her friends and loved ones. She is also likely to want to take photos in public places, where other people will see her using the camera.

A major social issue here is the impression of the user that the camera gives to others. Our target users are of a high socio-economic status. Perhaps they wish this to be reflected in the design of the camera? If they have paid a lot of money for the camera, then the camera's design should reflect this. In this way the camera can act as a 'badge' saying 'I'm a successful person'.

Product benefit: camera should confer the impression of high socio-economic status on the user.

As a target group that may be particularly image-conscious, it may also be that this group is concerned not only with socio-economic status, but also with cultural status. In other words, the design should give the impression that the user has not only affluence, but also a measure of good taste.

Product benefit: camera should confer the impression of high cultural status on the user.

The social context in which the camera is used may also have implications for the benefits that the camera should provide the user with. Presumably, these users will be mainly taking pictures of people that they know rather than of professional models. Models may be prepared to spend a considerable amount of time getting a pose just right and waiting whilst the photographer makes adjustments to the camera settings in order to get the shot just right. The chances are, however, that our users' friends and families will not be prepared to wait whilst she goes through such a rigmarole. This indicates that the camera's design should support her in taking pictures quickly.

Product benefit: camera should enable the user to take photos quickly.

Another socio-issue connected with camera use is the potential noise disturbance to others that taking a photograph may create. People may want to take photos in places – such as churches, concert halls, theatres, etc. – where disturbance would not be appreciated. Indeed, it may be a cause of embarrassment to the user and of annoyance to others. This may just as likely be an issue for this target group as for any other.

Product benefit: camera should be operable without disturbing others or embarrassing user.

Psycho-pleasure

An issue to consider here is the pleasure associated with taking the photos and the enjoyment of the outcome. Much of this may be associated with traditional usability issues – such as effectiveness and efficiency of performance. In this case, effectiveness may be associated with the quality of the photos, whilst efficiency may be about how easy it is to take the photos.

Again, it is important to consider the quality of photographs that these people will be after. As they are not professional photographers, they will probably be after good quality 'snaps' rather than professional-quality photographs. Whilst a professional photographer may be prepared to spend hours getting a shot just right, these users will probably not be prepared to spend a long time over a shot, no matter how good the outcome. The emphasis, then, should be on getting the shot right with minimal effort. This reinforces the point that arose under socio-pleasure, about the camera's design supporting quick operation.

Product benefit: camera should enable the user to take photos quickly.

Our users are young, dynamic, successful women. They probably don't have the time or inclination to sit down and spend a lot of time acquainting themselves with the camera before using it. Compared with many of the electronic and information technology products that people now use in their daily lives, a camera seems an inherently simple product. It seems unlikely – particularly for this user group – that people would be willing to tolerate a camera that they cannot pick up and use at the first attempt.

Product benefit: camera should be easy to use at the first attempt.

Ideo-pleasure

Choosing to use or buy a particular product over another may often repre-sent an 'ideo-decision' – a decision that reflects the tastes, values and aspirations of the purchaser. Here, those involved in the product creation process should be aware of the potential ideo-perspectives of the successful young women at whom the camera is targeted.

First, it seems sensible that the product should provide aesthetic pleasure to those for whom it is designed – those involved in the design process should, then, give some consideration to the issue of the sorts of aesthetics that this target group might like.

Product benefit: camera should give aesthetic pleasure.

Another issue affected by life in the postmodern era might be the extent to which the user would wish the product to be a reflection of her femininity. Twenty or thirty years ago overtly feminine designs may have been seen as 'patronising'. These attitudes are reflected, for example, in the feminist liter-ature of the 1970s, much of which viewed popular culture – including design – as a framework within which women were controlled and patronised by men (see Strinati 1995 for an overview). In an analysis of design from 1895 to 1980, Forty (1986) concluded that products tended to reflect stereotypes of men as plain, strong and assertive and of women as 'decorative', weak, delicate and sensitive. In such a climate, making an overtly 'feminine' product may have been seen as crass and demeaning, especially by a user group such as this one, which, presumably, would be well-educated and progressive.

Since then, however, paradigms in the West have shifted, both in the design world and in society as a whole. As women increasingly win equality and make socio-economic progress, many may also feel it less contentious to express their femininity through the products that they own and use. Indeed, the percentage of women working in design is ever increasing. These factors have led to a climate where 'feminine' design may be seen as positive and expressive, rather than patronising. Such attitudes are reflected in post-modern feminism, which tends to support the cultural representation of gender differences to a far greater extent than did the 1970s feminist move-ment (see Paglia 1995, for example).

Product benefit: camera should reflect the users' femininity.

Aside from 'personal' ideologies such as self-perceptions and aesthetic pref-erences, people's pleasure with products may be affected by their social, political and religious ideologies. An issue that seems likely to concern a young, well-educated group such as this is the environment.

Product benefit: camera should be environmentally 'safe'.

So, by considering the characteristics of the target group holistically, it has been possible to make some assumptions about the sorts of benefits that they may wish to gain from the product under consideration. In summary, the product specification derived from this example is listed in Figure 3.3.

In total, then, this product benefits specification contains twelve potential benefits against which the quality of proposed design solutions can be judged. Note that, in the table, no indication has been given of the pleasure category with which each of the benefits is associated. This is not important now. The four-pleasure framework was used as a means of structuring thought in order to arrive at the product benefits specification. However, once the specification has been decided, the framework has served its purpose. From now on the issue is about how to provide these benefits through the design of the product. In order to address this issue it is neces-sary to examine the links between these benefits and the properties of a product's design.

Specifying product design properties

The previous section looked at how to arrive at a product benefits specifica-tion through understanding the people for whom the product is to be

Camera should feel good in the hand
Camera should be easy to carry around
Camera fits well and comfortably against the face
Camera should be operable without causing damage to the users' fingernails
Camera should confer the impression of high socio-economic status on the user
Camera should confer the impression of high cultural status on the user
Camera should enable the user to take photos quickly
Camera should be operable without disturbing others or embarrassing user
Camera should be easy to use at the first attempt
Camera should give aesthetic pleasure
Camera should reflect the users' femininity
Camera should be environmentally 'safe'

Figure 3.3 Product benefits specification for a photo-camera designed for European women of a high socio-economic status aged between twenty-five and thirty-five years

designed and the context in which the product will be used. Having arrived at a product benefits specification, the design team is then faced with the question of how to create a design that will deliver these benefits – this is the subject of this part of the chapter.

First, a distinction will be made between a product's 'experiential' properties and its 'formal' properties, and the relationship of these properties to the benefits delivered by a product will be described. Both the experiential and formal properties will be discussed within the context of the various 'elements' – form, colour, materials, graphics, etc. – which go to make up a design. After this, the example of the photo camera case study will be continued, demonstrating how to move from a product benefits specification to a definition of a design in terms of its formal properties.

Formal and experiential product properties

A product can be defined by its properties. Properties can be either formal or experiential. Formal product properties are those that can be objectively measured or that have a clear and fairly unambiguous definition within the context of design. Experiential properties, meanwhile, are those that are defined in the context in which the product exists and of the views, attitudes and expectations of the people experiencing the product. For example, a motorcycle may have a top speed of 120 miles per hour. This is a formal property of the motorcycle – it can be measured objectively. Experientially, this top speed might be seen as either 'fast' or 'slow' depending on the person experiencing the motorcycle and the context in which the motorcycle exists. If the motorcycle was for use in competitive racing and the person experiencing the motorcycle was a racing rider, then he or she might regard the motorcycle as being slow – that is, he or she might regard the motorcycle as having the experiential property 'slow'. If, on the other hand, the motorcycle was intended for urban use and the person using the motorcycle was a commuter, then he or she might regard the motorcycle as being 'fast' – in other words, he or she might regard the motorcycle as having the experiential property 'fast'.

Similarly, a piece of furniture, such as a table, may have a number of patterns carved into it. Such patterning was very common during the Victorian era. Such a piece of furniture might be formally defined as 'ornate', as opposed, for example, to organic or geometric. Whilst 'ornate' is not a property that is measurable in the same way as the top speed of a motorcycle, it is, nevertheless, a property that would be recognised and interpreted fairly universally amongst designers. It may be thought of, then, as still being a formal property. Experientially, however, this ornateness may be interpreted in a variety of different ways depending on the person making the interpretation. Some people may see ornateness as being 'traditional', others may see it as 'fussy', others as 'dated', others as 'attractive' and still others as 'tacky'. So, the formal property 'ornate' could be linked to one or more of the experiential properties 'traditional', 'fussy', 'dated', attractive' or 'tacky' depending upon the person or people experiencing the table.

The first step in moving from a product benefits specification to a definition of product properties is to establish a link between the benefits that the product should deliver and the experiential properties required in order to deliver these benefits. These, in turn, can then be linked to the formal properties of the product. These links must be made in the light of the context of the product's existence and in the light of the views, attitudes and expectations of the people who will experience the product.

Consider, again, the example of the motorcycle. Imagine that the intended target group for the motorcycle was wealthy young to middle-aged men and women who wanted to use the motorcycle on weekends as an 'escape' from the mundanity of, for example, a sedentary professional life. It seems likely that 'excitement' would be one of the product benefits specified for a product such as this. Having defined 'excitement' as part of the product benefits specification, the following step would be to come to a decision about the experiential properties of a motorcycle that would be required in order to deliver the benefit 'excitement'. It might be, for example, that if the motorcycle were to provide excitement it would have to be seen as 'fast' and 'attractive' by those at whom it was targeted. 'Fast' and 'attractive' would, then, be defined as two of the experiential properties that the product must have.

The next step is to specify the formal properties associated with the experiential properties. This target group are not professional racing riders, yet they may still be expecting high performance. Perhaps, then, a top speed of 150 miles per hour would be seen as fast by this target group. So, the product benefit 'exciting' would be linked to the experiential property 'fast', which would be linked to the formal property 'top speed of 150 miles per hour'. The link between 'fast' and 'top speed of 150 miles per hour' is, conceptually at least, fairly straightforward. However, when considering the formal properties associated with the experiential property 'attractive' things might be more complex.

The complexity arises because an experiential property such as 'attractiveness' is likely to be associated with more than one formal property of the product. For example, 'attractiveness' might be associated with the colour of the motorcycle, the form language used and the materials and finishings used on the bike. So, for example, it might be decided that the motorcycle should be designed in a retro form, that the body should be fabricated from steel with chrome finishings and that the colour of the body should be black. Here, then, the experiential property 'attractive' would be associated with the formal properties 'retro form', 'steel body', 'chrome finishings' and 'black colour scheme'.

So, in this example, one aspect of the product – top speed – was associated with making the product fast, whilst four aspects of the product – form, body material, finishings and colour – were associated with making the product attractive. These aspects of the product – which, from here on, will be referred to as product 'elements' – can be thought of as the building blocks from which the product as a whole is created. The product's formal properties can be seen as particular manipulations of these elements. So, in the example of the motorcycle, the formal properties of the form, body material, finishings and colour were manipulated in order to make the product attractive. The next section is about elements.

Elements of product design

In this section six separate categories of product elements are identified. Elements can be thought of as the constituent parts of a design – the building blocks from which the overall design is created. Looking at each element in turn can help to identify opportunities to create particular experiential properties of a product through the manipulation of the formal properties of the product's elements.

Colour

Colour is an element that, in many different product domains and many different cultures, can evoke strong associations.

Fashion design is one area in which colour has fairly well-established associations. For example, in the West, the use of powder blue colouring in a baby's romper suit is likely to be seen as denoting masculinity. So, in the context of baby romper suits in the West, giving the element 'colour' the property 'powder blue' is likely to invoke associations with masculinity – 'powder blue' being the formal property of this element and 'masculine' being the experiential property. A baby wearing such a suit may be treated differently to one wearing a pink romper suit and the way in which he or she is treated may reflect the stereotypes that Western society holds about boys and girls. If the baby wearing the suit were a boy, and the mother and father of the child subscribed to 'traditional' views as regards gender roles, then they might see this as a positive affirmation of the baby's gender. On the other hand, if the child's parents hold 'progressive' views on the issue of gender roles, then they may wish to discourage others from treating the baby in a way that reinforces boy–girl stereotypes. Far from viewing such behaviour as a positive affirmation of their child's gender, they may view it as a negative influence on their child's social development.

In other countries, powder blue may not carry the same associations. If the romper suit was being sold in China, then the use of powder blue as a colour may not carry the same meanings as in the West – people would not necessarily associate the colour with masculinity, so the romper suit would no longer deliver any benefits (or penalties) connected with affirmation of the child's gender. In other words, in China, the formal property of a romper suit 'powder blue' may not be associated with the experiential property 'masculine'. Even in the West, powder blue will not necessarily always be associated with masculinity. In the 1980s, for example, there was a period when powder blue became popular as a colour for use in domestic appliances and soft furnishings. There was no suggestion that its use in this context in any way elicited associations with masculinity.

Automobile design is another area where colour seems to have fairly consistent associations. For example, in the West black seems to be a colour that is associated with status and sophistication: the limousines used to carry national dignitaries are almost invariably black; President Clinton is driven around in a black Cadillac whilst Tony Blair travels in a black Jaguar. Other European heads of state nearly always travel in black versions of cars manufactured in their own nations. Here, then, the formal property 'black' is associated with the experiential properties 'high status' and 'sophistication'.

Red, in the context of car design, tends to be associated with high performance and masculinity. Ferraris, for example, are nearly always red, whilst candy-apple red has become associated with rugged off-road vehicles, particularly in the USA where Jeep and other off-road manufacturers offer a number of models in this colour.

Colour can also bring practical benefits. In the early 1980s many car manufacturers offered models in a garish luminous orange colour. This was not a fashion statement, but rather was done in order that the vehicles could be seen easily at dusk and under the light from other vehicles. Indeed road builders and others working on the roads will often wear luminous orange colours in order that they can be seen in the dark.

A few years ago, the English soccer champions, Manchester United, stopped using their grey away strip after a heavy defeat by Southampton, a comparatively lowly team. They blamed their defeat on difficulties in seeing each other on the pitch, claiming that the grey colour was difficult to spot against the background of the stadium and the crowd. Whilst these claims attracted considerable derision, both from opposition supporters and from the sporting press, there may have been an element of truth in them. After removing the grey strip from their considerable range of away kits, Manchester United's away form improved and they were ultimately able to retain their championship title. Here, then, the formal property 'grey' was associated – according to Manchester United – with the experiential property 'invisible'!

Colour can also be associated with social belonging and ideology. Staying with football, for example, the wearing of team colours readily identifies supporters' allegiances – this can help to provide a strong feeling of group identity. Unfortunately, this feeling can sometimes escalate into feelings of hostility towards those wearing other colours – a sad fact attested to by the number of city centre pubs that refuse entry to people dressed in football colours.

Similarly, a number of manufacturers have made the use of colour central to their identity. For example, Philips, the electronics manufacturers, have their own shade of blue that they use in graphics. Billboards and stadium hoardings carrying the company name will almost invariably contain 'Philips Blue' either as a highlight colour or a background colour. McDonalds, the fast food chain, have a red and yellow colour scheme that helps make their packaging and advertising instantly recognisable. Indeed, they often use red and yellow as part of the colour scheme within their restaurants – helping them to establish a consistent, reliable image. Other manufacturers use the colour of their products themselves as a means of establishing brand identity. The consistent use of red by Ferrari is a case in point. So, here, the formal property 'red' has – through consistent use by a manufacturer of high-performance cars – become associated with the experiential property 'high performance'. Another example of the use of colour as a means of branding is that of IBM who, in the 1980s, used off-white product colouring as part of their identity.

Colour may also have a direct influence on a person's mood. For example, in the context of interior decoration, pastel greens and yellows are thought to be relaxing colours.

Colour is becoming a 'hot' issue amongst manufacturers of consumer durables. At one time, for example, it seemed to be common wisdom that products such as fridges, washing machines, dishwashers and freezers had to be white – indeed, such products were often referred to as 'white goods'. Increasingly, however, these products are being offered in an array of colours. At Domotechnica 1999 – Europe's largest domestic appliance trade fair – nearly all of the major 'white goods' manufacturers were offering products in a variety of colour choices. Manufacturers often see use of colour as a way of increasing customer base through increased choice. People have, for years, been offered colour choice in many types of product – soon they will have much more choice about the colours of the products that they put in their homes.

Form

Sometimes the form of a product will draw on metaphor in order to give a product experiential properties that lead to a particular product benefit. For example, manufacturers of irons draw on metaphor in order to make their products look speedy. Lower, sharper forms are considered to make the products look faster. This metaphor is based on speedboats. The formal property 'pointed' is, then, associated with the experiential property 'speedy'. The reason for employing such metaphor is an assumption that people find ironing an unpleasant task and will thus be attracted to products that give the impression that they will help to get the job done as quickly as possible. The lower, sharper forms give a speedier appearance. There may, however, be trade-offs here. For example, although the sharper irons may look faster, they may also appear less caring towards clothes. Perhaps the pointedness will appear somewhat aggressive, leading to fears of damage to the clothes being ironed. Again, the most appropriate form will depend on the attitudes and perceptions of the person experiencing the product. A person who values quality of ironing result over speed of ironing may be less inclined to desire speed as a product benefit, than someone who simply regards ironing as an unpleasant chore.

In the case of irons, the association between speed and shape is merely a perceived association. There seems no reason to assume that irons with sharper forms would actually enable a person to iron measurably more quickly than they could with less pointed irons. However, there will be other products – for example, cars and the speedboats on which the ironing metaphor is based – where the form will have a measurable effect on performance. In these cases, the aero- or hydro-dynamics of the products will be affected by their shape and it is, then, likely that sharper forms will provide measurable benefits in terms of speed.

In many product areas there are forms that have become iconic. In other words, there are certain forms that people within particular cultures have come to associate with particular products. Moving away from or challenging these icons can present either a disadvantage or an advantage depending upon the attitudes of the person at whom the product is aimed. For example, the designs of Philippe Starke challenge much conventional wisdom about what a number of products should look like. Consider, for example, Starke's juicer, designed for Alessi, shown in Figure 3.4. This has a very different appearance from conventional juicers – employing an extremely pointed form. Such 'extreme' designs attract attention and are likely to polarise the opinions of those who experience them. Whilst some may enjoy the humour and creativity inherent in looking at a product in such a new way, others may feel unsure about purchasing something that stands so far outside the conventional form of a particular product type. In the case of the product depicted, then, the formal property 'pointed' is associated with the experiential property 'extreme'.

'Extreme' is a term often applied to product forms that stand outside the ones usually associated with other products of the same or of a similar type. A designer may use form deliberately in order to make an 'extreme' statement – as with Starke's juicer – but sometimes, however, a product that stays on the market unchanged whilst the style of the rest of the products on the market moves on can become extreme almost by default. An example of a

Figure 3.4   Juicer designed by Philippe Starke for Alessi

product that has become extreme in this way is the Citroën 2CV. When this car was originally designed its form may well have reflected the conventional design wisdom of the time. However, the car was manufactured in its original form for so long that designs in the rest of the market moved further and further away from it, making it more and more extreme by comparison. The Citroën 2CV is illustrated in Figure 3.5.

Often, extreme forms will attract a niche market, made up of consumers who are attracted by the idea of having something a little bit different from what 'Joe or Josephine Average' has. However, sometimes extreme forms can prove so successful that they bring about sea changes in design in their product area. A product whose iconic form has changed over the past twenty years or so is the electric kettle. Go back about twenty years or so and the conventional form of a kettle was the 'dome with a handle and spout'. When the first 'jug kettle' came on the market this would have been seen as an extreme and innovative form. Now, however, the jug form has

Figure 3.5   Citroën 2CV

become the norm, at least in the Western world. The dome form may now be considered the more extreme of the two, perhaps being seen as a 'retro' statement – a form loaded with references to the past.

Probably, the reason for the prevalence of the jug kettle is that – in the context of kettles powered by an electric coil element – it is a more 'rational' design. Having the handle on the side and the spout on the front prevents any steam that may escape through the lid and spout from burning the user's hand – often a problem with dome kettles. Indeed, the form of the dome kettle derives from the days when kettles were nearly always heated on the stove. In this context, the dome was a rational design because it had a large under-surface in comparison to its overall volume – this helped to heat the water more quickly. A jug kettle has a small under-surface area in comparison to its overall volume. It would, then, be an inefficient design for a stove-heated kettle. Interestingly, the recent development of flat elements that cover the whole of the kettle's under-surface may now lead to a move back towards the dome design, because, as with stove heating, these will heat water more quickly if the ratio of the bottom surface to the overall volume of the kettle is increased. So, in the context of the form of flat-element kettles, the formal property 'domed' may be associated with the experiential property 'rational'.

In the case of the kettles, there are clear practical benefits associated with form – quicker heating on the stove for dome designs and usability and safety benefits for the jug design. These are opposed to the largely emotional benefits associated with the playfulness of the Starke juicer and the nostalgia of the Citroën 2CV. Indeed, form language has often been used to load products with cultural and other references in order to bring emotional benefits to those who experience them.

In the previous chapter, it was noted that design form could reflect the national psyche – in particular, this was embodied in the jet-age design forms that were popular in the USA in the 1950s and the space-age forms that were popular in the UK in the 1960s.

As well as the practical and emotional associations associated with form, form can also bring sensorial benefits. In the case of hand-held products, for example, form can play a major role in determining how pleasant or unpleasant the product is to hold. Again, this is an issue that has not escaped the attention of manufacturers. One area in which this can be seen is in the design of television and other types of remote controls. When remote controls first became popular, they were almost universally designed in a geometric rectangular form. Whilst these enabled the designer to lay out the buttons neatly across the surface of the remote, the rectangular form was not always very comfortable to hold. Recently, a number of remote controls have come on to the market that have a more organic form. These are easier to hold than conventionally shaped remote controls as they sit nicely in the palm. However, it is not simply a matter of being easier to hold – they also feel far more pleasant to the touch. The organic form gives a feeling of solidity and the lack of corners means that there are no sharp points to jab into the user's hand. Here, then, the formal property 'organic' is associated with the experiential property 'sensorially pleasing'.

The Wilkinson Sword Protector razor is another example of a hand-held product that, because of its form, is pleasant and easy to use. This product is designed in a well-balanced single organic form, with the handle weighted at its far end in order to counterbalance the weight of the head of the shaver where the blade is attached. This balance helps give the user a sense of control over his shave and the solid organic form radiates a feeling of solidity and quality.

A current trend in design is towards retro forms. These forms draw on past form icons, but are executed in such a way as to add an element of humour and expressiveness to the product. The philosophy underlying the use of such products is to offer new technology whilst at the same time making a reference to the era in which the product type was first conceived. Another trend is in the use of asymmetry as a means of adding style, perhaps even humour, to a product. An example of this is the clock manufactured by In-House, which is illustrated in Figure 3.6.

Figure 3.6   In-House clock

Product graphics

Graphics can be used on products for a variety of purposes. For example, graphics may be there to show how the product works, to give the product a particular look or feel, to advertise the product's functionality or to help give the product a particular style or ambience. Many software products use graphics to represent the fields, or 'soft-buttons' through which the user interacts with the product.

Apple computers were one of the pioneers of the graphical interface. For example, graphical icons are used on the Apple Macintosh in order to create the desktop metaphor through which the user interacts. Here, the metaphor is developed through the use of icons that look like, for example, files, folders, trash cans and printers. Because these graphical icons have these representational properties – in other words, because they look like the objects that the user may wish to manipulate in order to carry out a task – they help make the operating mechanism of the product usable. Here, then, the formal property 'representational' is associated with the experiential property 'usable'.

Because these graphical icons mirror everyday objects, they help to give the interface a non-technical look – something that can help make the computer less daunting for users who are unfamiliar with such high-tech products. Indeed, it was this interface style that proved central to the Apple Macintosh's success, as it made the machine approachable for a whole generation of users who might otherwise have been frightened by the idea of using computers.

On products where the interaction is via hardware – for example, buttons, knobs, dials and displays – graphics are often used as labels. Here issues such as the legibility of the graphics become important. The design team should have an understanding of the distance and conditions under which labels will be viewed and should ensure that the size, colour and font style of the graphics are such that the user will be able to read them from this distance. Consider, for example, using a car stereo whilst driving. Here the user will only be able to give cursory glances at the stereo, as most of the time his or her gaze will – hopefully! – be directed to what is happening on the road. A further difficulty arises because the position in which car stereos are typically installed – low down on or under the dashboard – means that the front panel of the stereo may also be poorly lit. This means that graphics, which may have seemed easily legible in the design studio, may not be nearly so legible in the real conditions of use. In this case, then, a number of formal properties such as the size of the graphics, the font style and the contrast with the product's body colour contribute to the experiential property 'legible'.

Standard ergonomics literature (e.g. McCormick and Sanders 1983) is a good source of information about what makes graphics legible. A general rule of thumb is that lower-case text will be more easily legible than upper case – as with lower case the shape of the word gives an additional visual cue – and that pictograms will be more legible than text – as pictograms lend themselves to quicker cognitive processing on the part of the viewer. However, in the case of pictograms, this rule of thumb comes with an important caveat – whether the object or action represented by the graphical symbol can be easily and quickly understood by the viewer.

Brigham conducted a study of the understandability of a selection of symbols that had been submitted to the International Electrotechnical Commission for approval for use on electronic products. His findings were rather shocking. For example, of a set of four different symbols demonstrating different timer functions, the meaning of two of these could not be guessed by any of the twenty-four interviewees that Brigham asked and the other two by only five of the twenty-four (21 per cent) (Brigham 1998: 11). Brigham concluded that many of the symbols that manufacturers put on products were not only difficult to understand, but were actually misleading, causing users to associate the symbols with different functions entirely. Some of the symbols evaluated by Brigham are illustrated in Figure 3.7 (the 'hit rate' represents the number and percentage of respondents who could guess the meaning of the symbols).

As well as denoting functions, pictograms can be used in order to give the

	Elapsed time	Remaining time	Programmable start	Programmable stop
Set 1 N = 24				
Hit rate	0 (0%)	0 (0%)	5 (21%)	5 (21%)
Set 2 N = 16				
Hit rate	9 (56%)	9 (56%)	6 (38%)	6 (38%)
Set 3 N = 48				
Hit rate	25 (50%)	27 (55%)	30 (64%)	28 (61%)

Figure 3.7   Selection of symbols (in this case timer symbols) evaluated by Brigham
Source: Brigham 1998: 11.

user feedback about the state of a product, or in order to prompt the user to take a particular action. Often the two issues will be linked – because a product is in a particular state, the user will be prompted to take an action. Perhaps, for example, the product will now be ready for the user to under-take a particular task, or perhaps the product's state requires the user to take some sort of corrective action. Again, the effectiveness of such pictograms is likely to be strongly linked to their representational properties. Where the state of a product has implications for user actions, Barnard and Marcel (1984) argue that pictograms which illustrate the proposed action will be more effective than those representing the product's state. This, they argue, is because a pictogram that gives a direct prompt towards action is less ambiguous than one that simply shows the product's state and leaves the user to surmise what the potential action consequences of this are. In this analysis, then, pictograms representing actions will be more effective as, providing the pictogram has been understood, the user will instantly know what the potentials or requirements for action are. However, with pictograms that only show the product's state, the user will have to make his or her own inferences about action potentials – this extra step is a possible source of error or delay.

Barnard and Marcel's theory is supported by Brigham (1998) who cites the example of two alternative icons designed to prompt the user of a domestic iron to refill the boiler when it becomes empty – these are illus-trated in Figure 3.8. The left pictogram conveys information about the state of the boiler – it tells the user that the boiler is empty. It is, then, left up to the user to decide what to do about this. The right pictogram, by contrast, represents the action of refilling the boiler. Provided that the user has under-stood this, it is instantly clear what he or she is being urged to do. So, it appears that the formal property 'representing action' is associated with the experiential property 'intuitive'.

A counter-argument to Brigham (1998) and Barnard and Marcel (1984) might be that, although action-representing pictograms may be more under-standable in terms of their implications for action, they have the disadvantage of taking control away from the user by being 'dictatorial' rather than 'infor-mative'. Considering again the pictograms in Figure 3.8, the right one is simply telling the user that they should refill the boiler. It is not actually

Figure 3.8   Alternative pictogram designs: 'boiler empty' (left) and 'fill boiler' (right)
Source: Brigham 1998: 12.

giving any explanation as to *why* he or she should refill it. The user does not know whether the boiler is totally empty, or whether it is just running a bit low. Indeed, it may not be prima-facie obvious that the reason for tipping water in is directly connected with the boiler at all – perhaps, for example, the user might think that there was some cooling system in the iron that relied on keeping the water level topped up. The left pictogram, by contrast, is simply telling the user that the boiler is empty – it is then up to the user if and when to react to this.

Perhaps the ideal solution is feedback that informs both about the state of the product and suggests possible user actions. Crozier (1994), for example, reports on the development of a display that used pictograms and text in combination to give both state information and action suggestions in the context of an in-car display. Prima facie, a disadvantage of such a display would appear to be that the user would require more time in order to digest the additional information inherent in providing both state information and suggestions rather than just one or the other. However, an empirical evaluation showed that drivers were able to respond to this combined display more quickly than to displays either showing only the state or only suggesting an action. This is consistent with previous findings that 'over-informing users' by supplying what is known as 'redundant' information is likely to improve performance (Crozier 1994).

An advantage of pictorial graphics over textual graphics can be that of international understandability. Clearly, there are a vast number of languages and dialects in the world and any product that is marketed internationally is likely to be used by people of differing linguistic backgrounds. This can give an advantage to using pictograms as opposed to text on internationally distributed products.

In the case of software-based interfaces, it may be possible to offer the user a choice from a selection of different languages through which they can interact with the product. This approach is becoming increasingly common with home entertainment products, such as televisions, video recorders and computer games. For example, many of the games available for consoles such as the Sega Megadrive, the Nintendo 64 and the Sony PlayStation include a choice of languages in their top-level menu options. Similarly, many menu-software packages – for example, many word-processing packages – now come in a variety of different language options.

Graphics can also have a major role to play in carrying the emotional values associated with products. For example, typefaces used on a product or a package can have a significant effect on how people perceive a product. Some typefaces seem more playful than others (e.g. Brush Script and Comic Sans), some seem more modern (e.g. Futura and Syntax), some seem more traditional (e.g. Palatino and New Century Schoolbook), whilst others seem more avant-garde (e.g. Bauhaus Medium and Braggadocio). These typefaces are illustrated, along with their suggested experiential properties, in Table 3.3.

Note that the experiential properties listed are merely suggestions for the purpose of illustrating a point – there is no suggestion that these are universally recognised qualities of each typeface.

Accepting, for the sake of illustration, that these typefaces do have the qualities suggested, it might then be possible to establish some links between the formal properties of the typefaces and the qualities – i.e. the experiential properties – that they radiate. For example, the typefaces that have been selected to illustrate the quality 'traditional' are both ones that have serifs, whereas the two chosen to illustrate the quality 'modern' lack serifs. Therefore, it might be suggested that the formal property 'serifs' is associated with the experiential property 'traditional', whereas the formal property 'no serifs' is associated with the experiential property 'modern'.

Looking at the 'playful' typefaces, the formal property that they have in common is an irregularity of form. So, a suggested association is that the formal property 'irregularity' might be associated with the experiential property 'playfulness'. The common formal property of the 'avant-garde' typefaces illustrated is that they are incomplete. For example, the 'B', 'a' and 'e' in the Bauhaus Medium typeface are open. Similarly, the 'B', 'a', 'g', 'd' and 'o' in the Braggadocio typeface are split. So, the formal property 'incomplete' might be associated with the experiential property 'avant-garde'.

Materials

The materials from which a product is fabricated can play a major role in determining how pleasurable – or displeasurable – a product is for those experiencing it. As an example, consider the environmental issues associated with materials. Care for the environment is an issue that has come to increased prominence since the beginning of the 1980s. Whilst before people may not have thought too much about the effect that their actions could

Table 3.3 Typefaces and their (suggested) qualities

Typeface	Quality (experiential property)
Comic Sans	Playful
Brush Script	Playful
Futura	Modern
Syntax	Modern
Palatino	Traditional
New Century Schoolbook	Traditional
Bauhaus Medium	Avant-garde
Braggadocio	Avant-garde

have on the environment, there is now an increasing awareness of these issues. Indeed, particularly in Western countries, care for the environment has now become a significant commercial issue. A significant percentage of consumers will take environmental issues into account when making purchase decisions and manufacturers who ignore this issue do so at their peril.

Many plastics are non-biodegradable and as such may be more harmful to the environment than noble materials such as metals, woods and ceramics. This has encouraged manufacturers to return to these traditional materials for many of their designs. For example, there is now a trend towards the increased use of metals in electrical goods. As an example, consider stereo systems. The 'shells' of these products – for example the speaker casings and the outside of the stereo itself – are increasingly fabricated from materials such as woods and metals. This is a reversal of a previous trend in which manufacturers had moved towards using plastics and away from noble materials. Plastics are cheap and – because they are so easily malleable – are comparatively easy to deal with in the fabrication process. These advantages were behind the earlier trend toward their use. However, the negative environmental associations with plastics have been powerful enough to persuade many customers to avoid products with a high plastics content and manufacturers have responded to this. Here, then, there appears to be a link between the formal property 'plastic' and the experiential property 'environmentally irresponsible'.

Nevertheless, the future for plastics need not be bleak. A number of biodegradable plastics have been developed and plastics have a number of other advantages that make their use very attractive to manufacturers and customers alike. Two of these – their cheapness and malleability – were mentioned in the previous paragraph. Other advantages include their lightness, that they are inert to most conditions (at least in the case of non-biodegradable plastics) and that they are easy to colour and texture.

One room in most Western households that has seen a marked increase in the presence of plastics since the Second World War is the kitchen. The pioneer of the use of plastics in the kitchen was American manufacturer Earl S. Tupper. In 1949 he patented a range of plastic food containers. These 'Tupperware' vessels were light, inexpensive and durable, and became extremely popular, especially in the USA and UK. Because they are inert to most environmental substances, plastics are, generally speaking, hygienic materials. This makes them ideal for use in products that come into contact with food (see Figure 3.9). In addition to food containers, food preparation products such as blenders, mixers, colanders and sieves are often made from plastic. This time, then, the formal property 'plastic' is associated with the experiential property 'hygienic'.

Plastics are also widely used in the food and beverage packaging industry where they have often taken the place of glass – for example, many bottles

Figure 3.9   Plastic is a hygienic material making it ideal for products that are used in food preparation

are now made from plastics. These are cheaper to manufacture than glass bottles, and they are also considerably lighter. This makes them far cheaper to transport and thus reduces transportation costs. Plastic is also far tougher than glass, making goods packaged in plastic less susceptible to damage and losses in transportation. These factors, in turn, mean that products packaged in plastics are likely to be cheaper at point of sale than those packaged in glass, making their use attractive to manufacturer and consumer alike.

However, there are some sectors of the food and beverage industry where plastics have not displaced glass as the material of choice. Perhaps the most clear example is the alcoholic beverage market. Here, the package – the bottle – tends to play a major role in the extent to which people enjoy the experience of consuming the beverage. Glass tends to have – in the perception of most people – more attractive tactile and visual properties than does plastic. As an example, consider the case of bottled beer. Many people enjoy drinking beer straight from the bottle. In this context glass may feel better in the hand – its coolness and weight being, perhaps, more sensorially pleasing than the lighter, warmer feelings associated with holding a plastic bottle. Furthermore, in the West at least, people tend to associate heavier materials with high quality (Macdonald 1998). Because of this, people may tend to associate a weightier bottle with a higher-quality drink. Indeed, this appears to be the assumption underpinning the continued dominance of glass in this sector of the market.

In the UK during the 1980s it became fashionable to drink beers straight from the bottle. Part of the reason for this seems to have been that drinking expensive premium lagers was seen as a status symbol. Drinking directly from the bottle made it clear that what was being consumed was indeed an expensive premium larger – if this were to be consumed from a glass, there might be a possibility that others in the bar might mistake the drink for an ordinary, cheap draught lager! Here, then, the design of the bottle was central to fulfilling the drink's role as a status symbol. Again, the higher quality associated with glass would have made plastic a non-viable material in this context. So, in this case, the formal property 'glass' was associated with the experiential property 'sophisticated'.

Indeed, materials are often used to add associated status to a product. The automotive industry was a forerunner in using materials in this way. For example, a particular model of a vehicle might come in a variety of versions, varying according to luxury levels. Whilst the lower-end versions might have had, for example, fabric seat covers and a plastic dashboard, the higher-end versions may have had, for example, leather seat covers and a walnut dashboard. The use of materials and finishings to differentiate on luxury level is now common throughout the consumer electronics industry. For example, manufacturers of electric shavers use a variety of different lacquers and inserts on their products in order to differentiate between luxury levels.

In the case of hand-held products such as shavers, the materials used will affect the tactile properties of the product. Soft-touch lacquers can be sensorially pleasing to the touch and are finding increasing use in hand-held products. An example of a low-cost product that uses soft lacquers is the Bic SoftTouch pen. This is largely fabricated from plastic, but has a soft-touch lacquer coating.

Whilst use of lacquers can greatly increase the perceived quality of a product, it is vital that the fabrication of lacquered products is executed to a very high standard and that the lacquers are tough enough to withstand a reasonable degree of impact. Where a lacquer is chipped away from a product, exposing a cheaper material underneath, this can not only look unpleasant, but may also give the impression of dishonesty. The owner of the product may feel cheated – he or she thought that the product purchased was high quality, but the quality now appears (literally) skin deep. Figure 3.10 illustrates the lacquered panel at the front of a pair of shoes. The lacquer on the panel gives the impression that the panel is made of metal, but has become chipped away to reveal cheap white plastic underneath. This not only ruins the aesthetic of the shoe, but may also leave the owner of the shoe with the impression that he or she has been cheated by a penny-pinching manufacturer, who may appear to have tried to deceive the customer for the sake of a minimal saving in manufacturing cost. In this case, then, the formal property 'lacquered' is associated with the experiential property 'dishonest'.

The way that materials are textured can also affect the way that a product is perceived. In the case of metals, for example, brushed, matt, knurled and gloss finishes are amongst the possibilities available. In the automobile industry, for example, gleaming, glossy finishes are used to give automobiles an air of class and sophistication. Gloss finishes may also have both real and perceived aerodynamic advantages over other styles of finishes. The use of glossy finishes was perhaps most spectacularly executed in the American car industry of the 1950s in which manufacturers offered their customers 'dream' cars decked in acres of gleaming chromium (Votolato 1998).

Gloss finishes can be more hygienic than other finish styles. Because they have no texturing, they are less susceptible to accumulating dirt and germs. This has led to their prominence on metals used in the bathroom and the kitchen. For example, taps and tap nozzles tend to be given gloss finishes, as do cutlery and other metal kitchen utensils. On the other hand, gloss finishes tend to show any dirt that does accumulate on them – even fingerprints can easily show up as greasy splodges – and matt and brushed finishes tend to be far more 'forgiving' in this respect. Figure 3.11 illustrates a kettle designed for Ikea. Here the metal body of the kettle is given a gloss finish. Wood has been used as the material for the handle and the lid and spout cover knobs. The use of noble materials – metal and wood – gives the product an air of timelessness and sophistication.

Figure 3.10　Shoe with chipped lacquering on the front panel

105

Figure 3.11   Kettle designed for Ikea

Materials and finishes can also play a significant role in determining how easy a product is to grip in the hand. For example, metal hand tools often have knurled handles so that they are less likely to slip from the user's grasp. An example is the knurled finish of the handle of the Stanley knife. In this case, it is the finish of the material that gives it its texture and hence its grippiness. Here, then, the formal property 'knurled' is associated with the experiential property 'grippy'.

Sometimes, however, it will be the properties of the material itself – rather than the properties of the finish – that will give grip. For example, the toothbrush illustrated in Figure 3.12 has a section on its handle fabricated from a rubbery material (possibly a silicone) that helps the user to grip the toothbrush even when the handle is wet.

Materials can also play a major role in safety. One of the major causes of death in house fires, for example, is from inhalation of highly toxic smoke. Some of the foams used in furniture are highly flammable and give off highly poisonous smoke. This has been the direct cause of many fatalities over the years, although recent legislation regulating the types of materials that are used in furniture has gone some way to addressing the problem. Returning to the automotive industry, legislation has also addressed the materials that can and cannot be used in vehicle interiors. For example, steering wheels must be made of materials that will, at least to some extent, cushion an impact in the event of an accident. This is in contrast to the unprotected metal steering wheels that were common until the mid-1970s.

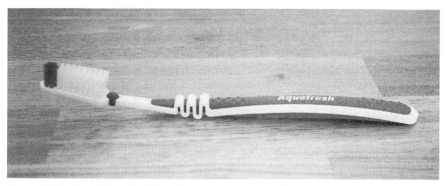

Figure 3.12   Toothbrush with rubbery implant in handle

Unusual and imaginative use of materials can make products particularly attention grabbing and make them particularly interesting to experience. For example, Japanese designer Shiro Kuramata specialises in the unusual use of metal in his furniture designs. He often uses metal latticing effects to create optical effects that make his furniture interesting to look at. These chairs seem to invite the viewer to sit in them in order to experience their unusual sensorial properties.

Avant-garde designers such as Kuramata have encouraged designers in major commercial studios to look again at the ways that materials are used. This has partly influenced the return to noble materials that was mentioned near the beginning of this section. It has also led to completely new uses for traditional materials. For example, Sharp now manufacture a stereo cased in ceramics.

Sound

Product sounds can give useful feedback about the state that a product is in. Broadly, product sounds can be considered in three categories according to their purpose: signals, navigating sounds and identity sounds (Jordan and Engelen 1998).

Signals are sounds that give cues about the state of the product: the sound of bubbling in a kettle indicates that the water has boiled; the roar of an engine indicates that a vehicle has started up; and the clicking of a switch indicates that a cassette tape has finished rewinding. These signals might be a cue for the user or others to react: switch the kettle off; put the car into gear and drive; or play the cassette tape. These examples refer to the natural or 'consequential' sounds that a product makes simply as a result of being in a particular state. It is also possible to design artificial or 'added' sounds that show a product's state. For example, a whistling sound to indicate that a kettle has boiled, a beeping sound to indicate that a truck is reversing, or a ringing sound to indicate that a telephone has an incoming call.

The examples given so far have been of one-off sounds, indicating that something has occurred. However, there are also signalling sounds that are continuously changing whilst the product is in use, indicating how the user of a product is progressing with some particular task. For example, the change in sound during a computer game as the player progresses through different levels, or faces particular dangers; or the change in pitch from a vehicle's engine, which indicates how quickly it is running and whether a gear change is required. This special category is known as navigating sounds. Again, these may be consequential, as with the engine, or added, as with the computer game.

The sounds associated with a product may have an effect on the general atmosphere surrounding the product. For example, products such as fan heaters create a constant hum. This may be relaxing or annoying, depending on the reaction of the person experiencing the product. Some products are designed for the express purpose of creating an atmosphere through their sound. For example, products that emit sounds aimed at relaxing the listener. The commercial implications of atmospheric sound are well-known within the retail trade. For example, shops often play music that they hope will entice people to buy particular products, or which will at least draw attention to particular products. For example, French music to draw attention to French wines and cheeses in a delicatessen, or rock music to create a lively, youthful atmosphere in a clothes shops.

Consequential or added sounds can also affect the identity of a product, or indeed the corporate identity of the manufacturer of such products. A number of automobile manufacturers, for example, have undertaken research programmes in order to optimise the sounds associated with the closing of car doors. If the doors close with a firm, solid sound, then people tend to associate this with quality. Hollow, tinny sounds, however, tend to give a negative impression of the overall quality of the car.

Similarly, some beer manufacturers have put a great deal of effort into the design of ring pull cans, in order to ensure that when the ring pull is opened, the 'hissing' sound gives the impression that the can contains high-quality beer.

The two examples given are of consequential sounds that can be deliberately manipulated in order to exude quality. However, the idea of communicating quality through sound applies equally to added sounds. For example, when switching on particular models of Apple Macintosh computers, the user hears a warm, deep, booming sound. Again, this conveys a feeling of robustness and quality.

Table 3.4 summarises the examples given above in terms of both the purpose and source (consequential or added) of the sound.

When designing sound into products, it is first necessary to decide what the purpose of the sound is to be and then to consider the properties of the sound that will be required in order to fulfil this purpose. These properties

Table 3.4 Examples of sounds sorted by purpose and source

	Source of sound	
Purpose of sound	*Consequential sounds*	*Added sounds*
Signals	• Boiling of water in kettle • Roar of engine when car starts • Click of switch when cassette finishes playing	• Whistle when kettle has boiled • Beeping of reversing truck • Ringing of telephone
Navigating sounds	• Pitch changes in vehicle engine	• Sounds in a computer game
Identity sounds	• Hum of fan heater	• Ambient sounds in stress-relief products

must then be realised either through manipulation of consequential sounds, or through the creation of added sounds.

One of the difficulties facing those involved in the sound design process is in knowing how to communicate with respect to sound issues. Discussing the purpose of the sound – giving a warning signal, relaxing the user, radiating quality – may not present any special challenges. However, discussing the formal properties of sound may prove more difficult. Although there are certainly formal properties of sound, such as pitch, timbre and volume, it can often prove difficult to find direct relationships between a sound's formal properties and its experiential properties. This point is particularly salient when the majority of participants in the sound design process are not sound professionals.

Typically, the creation of appropriate product sounds will require the expertise of a variety of professionals. In addition to sound designers and engineers, these may include product designers, market researchers, marketing personnel and product managers. For example, the product designer may consider the fit of the sound to the product's form and function, the marketing and market research personnel may consider the effect of sound on the marketing message behind the product, and the product manager may be concerned with logistical and cost issues associated with the creation of the sounds. Clearly, it may not be sensible to expect these people to understand or have a feel for the formal properties of sound.

In order to address potential communication difficulties, Jordan and Engelen (1998) report an approach whereby those involved in sound design can communicate using 'sound personality descriptors'.

In a brainstorm session, industrial designers generated a list of descriptors that could be used to describe the 'personalities' of sounds. These personalities were, in effect, the experiential properties of the sounds. This was followed by an evaluation session, in which the sounds associated with a selection of different products were evaluated by a team. In addition to the industrial designers, this team included a sound designer and a sound engineer. This session generated additional sound personality descriptors. In total around twenty-five pairs of sound personality – experiential property – descriptors were generated. Examples included: feminine versus masculine, strong versus weak, intense versus subtle, dirty versus clean, cold versus warm, sensual versus non-sensual, modern versus traditional. Participants in the evaluation session indicated that they found this approach to be meaningful and there was generally a high level of agreement about the descriptors that applied to a particular sound.

Having agreed on a set of experiential property descriptors, the team then set about creating sound briefings for two products – a sandwich maker and a solarium. A scenario-based approach was taken in which potential circumstances of product use were considered. These scenarios formed the context for developing the sound briefings.

The sandwich maker was used for preparing toasted sandwiches, which are mostly eaten as snacks. In the scenario depicted, the user was in the living room watching television, whilst the sandwich maker was in the kitchen. Three different sounds were specified, representing three different phases of the cooking of the sandwich. These corresponded to the sandwich being ready, the sandwich being browned and the sandwich becoming burned. It was specified that the personality of the sound should be different in each phase – starting as friendly and informative, but becoming increasingly stern and authoritarian if the user took no action to prevent the sandwich from burning.

Because the product was for food preparation, it was also decided that the sound's experiential properties should be modern, calm and clean. The form of the product was part geometric and part organic, so it was decided that the sound should also reflect this. In this case, then, the sound had three purposes: as a signal to inform the user of the state of the product; as a navigating sound to inform as to the degree to which the sandwich was toasted; and as an identity sound to indicate cleanliness, hygiene and efficiency.

The second sound design briefing was for an added sound to be included in a solarium. This product was for skin tanning; the lamp could tan half of the body at once. Because sitting in front of a sun lamp is a potentially boring activity, it was decided that the sound should provide the user with entertainment, relaxation and feedback – that is to say, that it should have the experiential properties 'entertaining', 'relaxing' and 'informative'. It was also decided that it should provide feedback as to how much longer the user had to remain in front of the lamp in order to tan to the chosen depth of colour.

The nature of the product suggested that a warm, calm and sensual sound would be appropriate. Meanwhile, the form of the product favoured a sound that gave a modern, geometric feel. In this case the sound was to be a navigation sound telling the user how long the program had left to run, and an atmospheric sound helping to entertain and relax the user.

The sound designer was briefed to create sounds in accordance with the properties summarised above. He then used his skill and judgement to produce sounds that exhibited these properties. These sounds were then evaluated by the project team in order to judge whether they fulfilled the briefings given.

In the case of the sandwich maker, the sound designer used electronic 'bleeps' as the medium for fulfilling the briefing. The frequency, volume, tone and pitch of these sounds increased as the state of the sandwich maker moved from 'sandwich toasted' through 'sandwich browned' to 'sandwich burning'. It was felt that the beeps gave the right sort of identity. The shrill pitch of the beeps and the precision and consistency of the timing gave, it was felt, the image of cleanliness, hygiene and efficiency. So, the formal properties associated with the pitch and rhythm of the sounds were associated with the experiential properties 'clean', 'hygienic' and 'efficient'. The manner in which the volume and frequency of the beeps increased was gradual but clearly audible. This, it was felt, gave the user a clear and authoritative – as opposed to hysterical – signal, as if the product were saying, 'The sandwich is now toasted, please come and remove it now.'

The sound designer approached the design of the solarium sounds by using the metaphor of a warm relaxing trip around the world. He created a number of musical sequences representing scenarios that ranged from sitting on the terrace of a French country house in the evening (e.g. the chirp of crickets, the clink of wine glasses) to lying on a deserted Caribbean beach (e.g. the sound of the sea, the sound of the breeze rustling the leaves in the palm trees). There were seven or eight different sequences that were programmed to play in a random order for as long as the solarium had been programmed to operate. The warm, rich sounds were achieved through the use of a deep bass sound that ran through the scenarios. It was felt that these sounds captured the essence of what the user would want to experience when using a solarium. They were luxurious and relaxing, giving the user a feeling of being pampered. The transition from one sequence to another was seamless because of the underpinning bass tones, and the unpredictable movement from one sequence to the next made the experience interesting for the user, even after much usage.

It is often difficult to pin down – in terms of formal properties – exactly what it is that makes a sound appealing. As an illustration of this point consider music. If a person were asked to describe their favourite types of music in terms of the formal properties of that music they would probably have trouble doing so. For example, it would probably be difficult for people to analyse why they preferred the Beatles or the Rolling Stones in terms of the formal properties of the music that they make. However, it may be easier for people to describe music in terms of its experiential properties – its personality. Table 3.5 gives some personality descriptors that, it is suggested, might be associated with the sounds of some of the best-known bands of the last thirty years or so. Note that these are merely suggestions given for the sake of giving a feel for what is meant by some of the sound personality descriptors – there is no suggestion that these descriptors represent a definitive critique of these bands' music!

Interaction design

This element refers to the way in which a product's interaction sequences and protocols are designed. It is an area that has received a great deal of attention from traditional human-factors approaches, from early studies on 'knobs and dials' to contemporary work that focuses largely on interaction with software-based interfaces.

Whatever the interface style, an important issue to consider is that of functional grouping. This is the basic principle dictating that controls which operate similar functions should, in some way, be grouped together. Usually, this means putting controls for similar things within close physical proximity of each other. Consider, for example, a stereo system containing a CD player, a cassette player and a radio. Usually, the controls for operating the CD player will be clustered together with one another, as will the controls for the cassette player and the controls for the radio. If these controls were to be 'mixed together' – for example, if some of the CD controls were

Table 3.5 Suggestions for sound personality descriptors associated with the music of some leading bands

Band	Examples of music (singles)	Experiential properties (personality)
Frankie Goes to Hollywood	'Relax'; 'Two Tribes'; 'Welcome to the Pleasuredome'	Clean; Warm; Energetic
Motorhead	'Ace of Spades'; 'Motorhead'; 'Overkill'	Dirty; Dynamic; Grinding
Sex Pistols	'Anarchy in the UK'; 'God Save the Queen'; 'Pretty Vacant'	Aggressive; Angry; Vibrant;
Abba	'Dancing Queen'; 'Thank you for the Music'; 'Super Trouper'	Cheerful; Bouncy; Energetic

clustered with some of the cassette player controls, then this might create confusion.

The same issue holds for menu-based interfaces. In this case an important issue is which commands go on which menu. Generally speaking, similar commands should be placed on the same menu and the menu heading should give a description of what it is that these commands have in common. This book is being typed on a word processor that contains fourteen separate scroll-down menus as well as ten separate toolbars. One of these menus contains, amongst other commands, the commands for the spell checker, the thesaurus, counting the number of words in the document, creating address labels and for customising the way in which the user interacts with the word processor. The menu containing these commands is headed 'Tools'. It seems, then, that the assumption behind the functional grouping used is that the user will regard these functions as having something in common – that they are all tools.

Functional grouping is an issue that comes under the umbrella of the wider interface design principle of 'consistency'. This principle states that the interface should be designed such that similar tasks are done in similar ways (Jordan 1998). The issue that those designing interfaces should consider here is which tasks – in the mind of the user – will be considered similar. Imagine, for example, a word processor that contains a function which enables users to put text into bold lettering and another to put text into italic lettering. Imagine that the protocol for putting text into bold lettering was:

1   highlight text to be put into bold lettering;
2   open the menu headed 'Format';
3   select the command 'Bold'.

Now imagine that the protocol for putting text into italic lettering were:

1   highlight text to be put into italic lettering;
2   open the menu headed 'Format';
3   select the command 'Italic'.

The protocol for these two tasks is similar – both involve highlighting text, opening the menu headed 'Format' and selecting a command from this menu. Provided that users regard these tasks as similar, then the interface will be consistent with respect to these tasks. In this case, it seems, prima facie, a reasonable assumption that users will, indeed, see these tasks as similar. After all, they are both tasks that have to do with changing the appearance of – 'formatting' – text.

Note, though, that 'consistency' is an experiential property of an interface design – it is a property that exists only in relation to a user's perception of which tasks are similar. So, the formal property 'command for putting text into bold lettering on the same menu as the command for putting text into italic lettering' is associated with the experiential property 'consistency'. Consider again the earlier example concerning the commands placed on a menu headed 'Tools'. Consider, specifically, two of those commands: 'Word Count' – for counting the number of words in a document; and 'Customise' – for customising the way in which the user interacts with the word processor. If the user were to regard these as being similar tasks, then placing them on the same menu would be necessary in order to make the interface consistent from the point of view of that user. On the other hand, if the user did not regard them as being similar, then placing them on different menus would not be an inconsistency. Indeed, having them on separate menus may seem logical to the user – they are different types of task so he or she might expect them to be activated via different protocols.

Consistency is a property of interface design that is recognised as being central to product usability. Some other, well-established principles of usable interface design are listed in Table 3.6.

Clearly, interface design is likely to have a major influence on how usable a product is. However, usability is not the only issue that will be affected by the design of an interface and the interaction protocols through which it is operated. In the case of computer-based interfaces these issues can also have an effect on the extent to which an interface 'engages' a user – the extent to which a user feels immersed in the virtual world of the interface. This issue has been tackled at some length by Brenda Laurel in her book, *Computers as Theatre* (1991).

Laurel believes that those designing software packages should view the computer as a medium through which interaction and collaboration between the human and the computer can take place. Both human and computer exist together in this virtual world where all that is represented is all that is important. Laurel maintains that the way in which this virtual world is represented should make explicit the assumptions that are made for both the human and the computer with respect to the actions that can occur within this world and the limitations on behaviour of human and computer. In this sense, then, the virtual world of the computer can be seen as a stage and the computer and the human as actors in a play. The plot of the play develops within the boundaries and rules of the 'world' in which the actions are being carried out.

Different plays have different rules governing their world. In *Peter Pan*, for example, the hero of the play can fly. This is not so in, for example, *Hamlet* or *Romeo and Juliet*. In other Shakespeare plays, such as *Julius Caesar* and *Macbeth*, there are characters who can predict the future – such as the witches in *Macbeth*. Other plays include characters who have special supernatural or magical powers, for example the sorcerer in *The Sorcerer's Apprentice* or the Good Witch of the East in *The Wizard of Oz*. Sometimes

Table 3.6 Summary of design properties of usable interfaces

Interface property	Definition
Consistency	Similar tasks are done in similar ways
Compatibility	Method of operation is compatible with users' expectations based on their knowledge of other types of products and of the outside world
Consideration of user resources	Method of operation takes into account the demands placed on the users' resources during interaction
Feedback	Actions taken by the user are acknowledged and a meaningful indication is given about the results of these actions
Error prevention and recovery	The likelihood of user error is minimised and if errors do occur they can be recovered from quickly and easily
User control	The extent to which the user has control over the actions taken by the product and the state of the product is maximised
Visual clarity	Information displayed can be read quickly and easily without causing confusion
Prioritisation of functionality and information	The most important functionality and information is easily accessible to the user
Appropriate transfer of technology	Appropriate use of technology developed in other contexts to enhance the usability of a product
Explicitness	Cues are given as to a product's functionality and method of operation

Source: From Jordan (1998).

there are special rules governing how characters can be killed or brought to life: pouring water on the Wicked Witch of the West in *The Wizard of Oz*; driving a stake through the heart of a vampire in *Dracula*; and Snow White being kissed by the Handsome Prince. Characters in these plays are presented with opportunities and obstacles that arise from these rules.

Similarly, when users step into the virtual world of a computer program, they are faced with a set of rules and boundaries that will govern the opportunities and limitations in terms of the actions available to them. For them to succeed in this world, they need to have a clear understanding of what these boundaries are. Central to users succeeding and enjoying interactions is that users become 'engaged' in the virtual world. The world of the play can engross an audience to such an extent that, in a sense, they 'believe' that what they are seeing on the stage is actually happening. This level of engagement is advantageous to the playwright as the audience may believe anything that seems remotely believable within the context of the world's rules. An understanding of these rules also helps the audience to predict what might happen next. However, by the same token, anything that happens in the play which violates these rules is likely to either render the action unbelievable or to create a state of confusion in the viewer as to what the rules of the world of the play are.

For example, in a Superman movie, audiences have no difficulty in accepting a scene whereby Superman saves a character falling from a building by flying towards him or her and catching him or her before impact with the ground. Indeed, those who are familiar with Superman and know the rules associated with his world would probably be able to predict that this would happen. If, however, something like this happened in a James Bond movie, it would be likely to throw the audience into a state of confusion and incredulity – it would have violated the rules that govern the world of these movies and would thus be totally unpredictable. The rules governing the world of James Bond movies are, in this sense, more a reflection of the rules of the real world than are the rules governing Superman movies. However, whilst they are sitting in the cinema, a Superman audience can unquestioningly 'believe' in the film provided that they are engaged in it and that consistency is maintained within the fantasy world.

Engagement and consistency can be equally valuable in human–computer interaction. If people become immersed in the world of the program that they are using and understand the rules that apply there, then they are likely to find the experience more enjoyable and are likely to be able to complete tasks more successfully. The use of metaphor in an interface is a good example of this. For example, the desktop metaphor used in the Apple Macintosh computer brings users into the world of a virtual office and has a fairly consistent set of rules governing the actions that are required in order to complete tasks in that world, even though the rules do not exactly mirror the real world.

In the world of the Apple desktop, dragging an icon on to another icon results in the item represented by the dragged icon being 'put' into the object represented by the second icon. For example, dragging a pictogram that looks like a piece of paper (a file icon) and placing it on top of a pictogram that looks like a trash can (a folder where files to be discarded are stored) results in a file being placed in a folder of items to be discarded. As long as the user is engaged in this world and understands the rules that operate in it, he or she will be able to predict, for example, that dragging another file icon on to, for example, an icon representing the hard disk will result in a file being moved to the hard disk. This contrasts with the older command line interfaces in which users type alphanumeric strings in order to carry out actions. This is likely to be less engaging as the users are not being given the chance to enter a new world, but are simply learning codes which enable them to interact with technology – more akin to learning a new language.

Laurel suggests that a means of increasing the level of engagement in a human–computer interaction is via the use of agents in an interface. An agent is a character within the virtual world who is there to interact with the user in order to, for example, achieve a particular goal. For example, the 'footballers' within a football computer game are agents that enable the user to try and 'score a goal' against the agents that represent the 'other team'.

In the design of a hypertext-based information system, Clarke *et al.* (1995) employed agents in an attempt to make the system more engaging and pleasurable for users. Appropriately, the information retrieval system was a database for use by film and theatre studies students and contained archived information relating to plays and playwrights. Clarke *et al.* decided that in order to help users navigate the information system they would employ a library metaphor – the user, then, should feel that when accessing the information retrieval system he or she had entered a library where information relating to plays was stored. Within this virtual library were three sorts of 'librarians' – agents who helped the users to access information that they were searching for or which might be of interest to them. One librarian was a playwright, one a producer and the other a critic.

When the user accessed information about a particular play the agents would appear and offer the user further information. Clarke *et al.* give the example of information pertaining to the play, *Saved*, written by Edward Bond. This play gained notoriety when it was first shown in the 1960s because it depicted violent youths. Indeed, it was banned from public performance, but became critically acclaimed when the ban was removed a few years later. As users read through the text pertaining to the play the agent icons appear at various times and places, leading the user to supplementary information that is relevant to what they are reading. For example, when the user accesses a page mentioning the ban on the play, an icon representing drama critic William Gaskell appears. By clicking on this icon the user will be presented with views expressed by this critic at the time. In addition 'William Gaskell' will also suggest links that the user might like to follow, perhaps giving other critics' views on the play or leading to information about other plays that proved controversial. Similarly, when the user accesses a page mentioning the playwright an icon representing Edward Bond appears. Clicking on this gives access to a page outlining Bond's thoughts on his own play and his reaction to the ban.

Clarke *et al.* extended the library metaphor further by providing the user with an on-screen notebook in which they can 'jot down' notes that can be printed out later. This allows users to copy text from the system and to paste it into the notebook. This differs from the usual Macintosh 'clipboard' system in which previously pasted text is overwritten by the latest text to be pasted, thus preventing the user from accumulating a number of different text fragments.

In a user evaluation, Clarke *et al.*'s system was compared with another version, which was built according to the usual principles of hypertext – accessing new screens through clicking on highlighted links. The information systems were evaluated according to three criteria – usability, engagement and overall pleasurability. Each participant in the evaluation used both systems in order to access information. Whilst using the system they thought aloud about what they were doing and how they were feeling. An analysis of their verbalisations was used as a basis for indicating the level of engagement with each of the systems. For example, if participants continually referred to the computer throughout their interactions with the system then it was concluded that they were not engaged in the virtual world of the system. If, conversely, they referred to themselves and their actions and to the world of the system, then it was concluded that they were engaged during usage. For example, a verbalisation such as 'I am just going to see what William Gaskell has to say about the play' would have been interpreted as a sign that the user was engaged. Afterwards they completed a usability questionnaire – the System Usability Scale (Brooke 1996) – and this was taken as a measure of usability. They then answered questions pertaining to which of the systems they preferred and which was the more pleasurable to interact with.

As expected, the analysis indicated that Clarke *et al.*'s system was far more engaging that the traditionally designed system. However, the scores given on the System Usability Scale showed no advantage for their system – mean scores for both systems being identical. Nevertheless, when asked which system was the more pleasurable to use, Clarke *et al.*'s system was unanimously preferred.

There are, perhaps, three major points to note about the outcome of the evaluation. The first is the indication of a strong link between the level of engagement experienced by participants and the level of enjoyment which they experience with the interface. This suggests that Laurel is correct in her premise that engagement is key to the enjoyment of users when interacting with computers. Aside from the outcomes of this study, strong support for Laurel's assertion would appear to come from the huge enjoyment that many people appear to gain from playing computer games. Here the user is taken into another world and given an aim of some sort – perhaps to find treasure, kill some baddies or win some sporting event. Many people can spend hours enjoying these games and would almost certainly not think of themselves as performing a task with a computer, but rather as completing some 'mission' within the virtual world of the game.

Second, that Clarke *et al.*'s system was seen as more engaging than the traditional interface would seem to support the suggestion that the use of agents in an interface is likely to make that interface more engaging to users. Here, then, the formal property 'use of agents' is associated with the experiential property 'engaging'. Again, computer games provide another example of this. With these games the player often takes on the role of a character through which he or she interacts with the world of the game. Presumably, this is a major factor in making such games so engaging.

Third, the results were further evidence for one of the major premises behind 'new human factors' – the idea that usability is only one of many issues affecting the overall quality of the person–product relationship. The results indicated that although users felt that there was nothing to choose between the systems in terms of usability, one of the systems appeared vastly more pleasurable to use than the other.

So, in summary, the formal properties of a product can be thought of as existing in the context of the various product elements and it is these formal properties that contribute to the experiential properties of a product. Before returning to the case study of the photo-camera, consider one more example. In this case consider the design of a piece of furniture – a desk. Imagine, for example, an office desk with a surface that measured, say, three metres by one-and-a-half metres. These dimensions – a part of the desk's form – are part of the formal properties of the desk. They can be objectively measured and will not alter in varying contexts or when different people experience the desk.

Experientially, if the desk were for use by one person only then the desk might be considered large – it would, in other words, have the experiential property 'large'. However, if the desk were for use by, say, six office workers at once, for example in the context of an open-plan office, then the desk might be considered 'small'. Similarly, different people in different cultures and contexts are likely to bring different perceptions to a product that will, in turn, alter an element's experiential properties. For example, if the six people working at the desk were telephone operators in Tokyo – where office conditions tend to be very cramped due to the prohibitive costs associated with hiring office space – then they might regard the desk as being 'large'. On the other hand, if the desk were for use by one person, but that person was the multi-millionaire chairperson of a Texan oil firm, then – if the stereotypes shown in soap operas such as *Dallas* and *Dynasty* are to be believed – he or she might regard the desk as being 'small'.

Table 3.7 gives examples of what the experiential properties of the surface dimensions of such a desk might be, depending on the people experiencing the desk and the context in which the desk is experienced. Some more user and context-of-use examples have been added to those discussed in the previous paragraph.

Table 3.7 Formal and experiential properties of the surface dimensions of a desk ('surface dimensions' would be an aspect of the element 'form')

Context	People	Formal property	Experiential property
One user	Texan oil chairperson	3m × 1.5m	Small
One user	London-based accountant	3m × 1.5m	Large
Six users	Tokyo-based telephone operators	3m × 1.5m	Large
Six users	London-based accountants	3m × 1.5m	Small
Two users	Amsterdam-based academics	3m × 1.5m	Medium

Case study part 2: deriving a property specification for a photo-camera

In part 1 of this case study, earlier in this chapter, a product benefits specification was derived for a photo-camera designed for European women of a high socio-economic status aged between twenty-five and thirty-five years. The product benefits specification is listed again in Figure 3.13. In part 2 of this case study a product property specification will be derived from this product benefits specification.

Each of these product benefits will now be considered in turn and suggestions made as to the sorts of product properties – both experiential and formal – which could be associated with such benefits.

Camera should feel good in the hand
Camera should be easy to carry around
Camera fits well and comfortably against the face
Camera should be operable without causing damage to the users' fingernails
Camera should confer the impression of high socio-economic status on the user
Camera should confer the impression of high cultural status on the user
Camera should enable the user to take photos quickly
Camera should be operable without disturbing others or embarrassing user
Camera should be easy to use at the first attempt
Camera should give aesthetic pleasure
Camera should reflect the users' femininity
Camera should be environmentally 'safe'

Figure 3.13 Product benefits specification for a photo-camera designed for European women of a high socio-economic status aged between twenty-five and thirty-five years

Camera should feel good in the hand

To deliver this benefit, it is important that the camera fits the hand well – in this case the adult female hand – and that it is nicely balanced in terms of weight distribution. In order to maximise the tactile pleasure associated with the camera, it should be light enough so that it is not too heavy to hold, yet heavy enough that it gives a feeling of solidity and quality. The materials used for the camera should also contribute to a sensorially pleasing feeling when held. These are experiential properties of the product that are required in order to fulfil the specified benefit. Now consider what formal properties might be associated with these experiential properties.

Fitting the hand well is likely to be associated with the form of the product. In this case the dimensions of the product are likely to be a major factor. The form of the product is also likely to effect whether the camera is well-balanced in terms of weight distribution. Formal specification of the product's weight will influence whether it is light enough to carry and heavy enough to give a feeling of quality. Again, though, which 'formal' weight is associated with the experiential properties 'light enough to hold' and 'solidity and quality' will depend on both the physical and cultural characteristics of the people for whom the product is designed. The formal properties of the material which are likely to be of influence in terms of making the product sensorially pleasing will probably include the type of material used and the texturing on the material surface.

Any moving parts should also work in a way that provides tactile satisfaction to the user. For example, the lens cover should slide open in a satisfying way – experientially it should not be too stiff to slide, nor too loose. Similarly, the level of resistance offered by the shutter button when taking a shot should be appropriate and should give a satisfying 'click'. These experiential properties of the camera are likely to be associated with the formal properties of the interaction design and can be defined in terms of, for example, the force required to depress the shutter button and the force required to slide open the lens cover.

Camera should be easy to carry around

This is likely to be a function of the weight and dimensions of the product which, experientially should be 'not too heavy' and 'not too big'. Maybe the person using the camera carries a handbag, so the dimensions of the camera might be specified such that it will fit in a handbag. Maybe she doesn't carry a handbag. The camera should also fit well into coat pockets, trouser pockets, jacket pockets etc. The shape of the camera should be designed such that it causes no physical discomfort when carried in a pocket. For example, it shouldn't have any edges that jab against the body.

Fits well and comfortably against the face

It is important that the camera is shaped so as not to jab into the face when being used. The materials used in the camera should also feel comfortable against the face.

Camera should be operable without causing damage to the users' fingernails

Small buttons and catches, which users may use their fingernails to operate, should be designed such that they do not offer a level of resistance that is likely to cause long fingernails to get broken. Alternatively, the design should avoid buttons and catches that will require fingernail operation. Again, then, it is the properties of the interaction design which are likely to be important here.

Camera should confer the impression of high socio-economic status on the user

This is likely to be associated with experiential properties such as 'exclusive looking' and 'expensive looking'. Because it has also been stated that the camera should confer an impression of high cultural status, it is important that the impression of high socio-economic status is not conferred in a way that would undermine this. This might mean, for example, that the design should not look 'flashy' or 'vulgar'. The formal properties of the design elements associated with the aesthetic appearance of the camera are likely to have a significant influence here. For example, perhaps the materials used should be noble materials. This might mean, for example, using metal rather than plastic for the body of the camera and, perhaps, encasing the body in a leather wrap. The form of the product is also likely to be important. For example, a postmodern form might give more of an appearance of exclusivity than would a traditional iconic camera form. The style of graphics used on the camera may also have an influence here – perhaps overly playful graphics might look cheap. This might mean going for textual characters with a regular, geometric form.

Camera should confer the impression of high cultural status on the user

This might be associated with experiential properties of the design such as 'sophisticated looking' and, again, 'exclusive looking'. Again, these will be associated with the formal properties of the elements that pertain to the aesthetics of the design.

Camera should enable the user to take photos quickly

A possible solution here is to make the camera a 'point and shoot' – one that the user simply takes out of her pocket, points at the object or people that she wishes to photograph and presses the button. This suggests that features such as auto-focus, auto-iris, auto-flash, auto-wind, and auto-just-about-everything-else will be appreciated.

Camera should be operable without disturbing others or embarrassing user

One issue that may be of importance here is the sound that the camera makes in operation. Many auto-functions rely on small motors to drive them. Sometimes the noise generated by these can be embarrassing. A colleague inadvertently terrified fellow passengers on a flight from Amsterdam to New York. He was taking pictures through the window of the aeroplane when he came to the end of his film and the camera started to rewind the film automatically. Unfortunately, the sound of the film rewinding was so loud and grating that the passengers in his and the neighbouring rows of seats thought that something was horribly wrong with one of the aircraft engines. It was only by pointing to the camera and apologising that he was able to calm the panicked faces around him!

Camera should be easy to use at the first attempt

It is important, then, that the camera is 'guessable'. 'Compatibility' is an experiential property that tends to be associated with guessability. This might mean that, in terms of the formal properties of its interaction design, the operation protocols for the camera should be those that are typically associated with camera operation. So, it might not be appropriate to introduce a novel interaction design in this camera.

Camera should give aesthetic pleasure

What would this group's idea of a beautiful design form be? Based on both design precedent (Tambini 1996) and on empirical evidence (Hofmeester *et al.* 1996), it seems that there is a preference amongst younger women for organic, postmodern forms as compared to, for example, more geometric, modern forms. In this case, then, the experiential property 'beautiful' might be associated with 'organic' and 'postmodern'.

Camera should reflect the users' femininity

An experiential property that is often associated with femininity is that of 'elegance'. Indeed, when Canon designed their Elph camera – a camera specifically designed to appeal to young women – this was one of the main experiential properties that they were trying to communicate (Shiotani 1997). In this case they felt that this was associated with the form of the camera and the materials and finishes used for the camera. The formal properties of the dimensions of the camera made it compact and slim. Meanwhile, the body of the camera was fabricated from stainless steel that was finished with a matt texturing. In this case, then, the dimension specification, the use of stainless steel and the matt finish were all formal properties that were associated with the formal property 'elegance'. The Elph is illustrated in Figure 3.14. It is interesting to note that the camera design achieves an elegant, feminine appearance despite the use of a geometric form language. This camera has been well-received in the market and won a Gold Award from the Industrial Designers Society of America.

Camera should be environmentally 'safe'

This was also an issue for Canon when they designed the Elph – indeed, it was another reason behind the choice of stainless steel as the body material (Shiotani 1997). There are now also a number of biodegradable plastics available that are more environmentally friendly than more traditional plastics. The processes used in the manufacture of the camera can also be an issue here. Again, this was something which Canon considered, avoiding finishes that would have required the use of certain environmentally damaging chemicals.

Evaluating designs for pleasurability

Having generated some design ideas and concepts, the next step is to evaluate these to investigate as to whether they are likely to lead to a product that meets the criteria laid out in the product benefits specification. In this next section evaluation is discussed. First, the issue of prototyping will be discussed and a number of different sorts of prototype will be described. This is followed by a discussion about setting the criteria by which the quality of a design can be judged.

Prototyping

In order to evaluate a design idea, it is necessary to first embody that idea in a prototype. In their most basic form, prototypes may simply be lists of product benefits or product properties. At the other extreme, prototypes

Figure 3.14   Canon Elph camera

may be fully operational representations of a product. Finished, manufactured products may also be evaluated. A number of different types of prototype, of differing levels of realism and sophistication, are discussed below.

Product benefits specification

The product benefits specification can, in itself, be thought of as a prototype. The benefits contained in the specification can be evaluated as to their suitability in the context of the product and the people for whom it was designed. These benefits will usually be presented in the form of a list – perhaps with extended written descriptions of each benefit. This list can then be evaluated. Perhaps people representative of those for whom the product is designed might be asked whether they would appreciate a product that delivered the suggested benefits. Alternatively, human-factors experts might be asked to pass judgement on the suitability of such benefits or the benefits might be evaluated against some pre-set criteria.

Product property specification

The product specification can also be evaluated. Again, this might be presented in the form of a list giving the suggested experiential and/or formal properties of a product. Participants in an evaluation might be asked, for example, whether they felt that a product which exhibited the experiential properties listed would deliver the benefits listed in the product benefits specification. They might also be asked whether the suggested

125

formal properties of the product would, in the context of the people for whom the product is designed, give the desired experiential properties.

Product property specifications need not only be given in the form of written lists or verbal descriptions. It may also be possible to give visual indications of the product's properties. The Focus Board is a tool – developed at the Glasgow School of Art – which is used for visualising the experiential properties of a design. Focus Boards are not visual prototypes – they do not give a literal visual representation of a design, but rather they give a feel for the general ambience associated with a particular experiential property.

Visual tools, such as Focus Boards, are widely used in nearly all major design studios (Jordan and Macdonald 1998). It is often felt that they provide a far more effective medium of communication than the written or spoken word, particularly because design is largely a visual-based activity. Figure 3.15 illustrates another such tool – the Visual Communication Board. This is used as a means of illustrating the 'visual world' to which a product belongs. In this case the board illustrates a number of different types of masculinity. The board was used as input to a product designed for a variety of male target groups.

Visual prototypes

These are simply visual representations of a product. They could be paper-based sketches or drawings, or on-screen representations created using software-based drawing or illustration tools or using computer-aided design tools. They may be supplemented with animated, written or oral descriptions of the product's functionality or interaction mechanisms. Initial concepts can be shown to people or evaluated against predetermined criteria, allowing judgements to be made about the aesthetic and functional qualities of the product. It can also be meaningful to ask users about their perceptions of the ergonomic aspects of the product, such as how easy to use they expect the product will be. Clearly, however, users will not actually be able to interact with a visual prototype, so judgements will reflect perceived usability rather than actual interaction performance.

Visual prototypes may be most effective in situations where acceptance of a product is likely to be highly dependent on its form and functionality rather than the effectiveness and efficiency of use. Consider, for example, the design of household lights and lamps. In this case, user satisfaction with the product may be highly dependent on the product's form, rather than on the interaction design. After all, lamps can generally be operated without too much difficulty as they tend to have few functions – usually just an on/off switch or perhaps an additional dimmer function.

Visual prototypes may be less effective when a person's experience of a product has a significant tactile element. Household lights are mainly expe-

Figure 3.15 Visual Communication Board illustrating different
types of masculine atmosphere

Source: Macdonald and Jordan 1998: 553.

rienced visually, again making them a good example of a product that can
be effectively evaluated via visual prototypes. Similarly, people's experience
of products such as television sets, video cassette recorders and stereo
systems will be largely dependent on their visual appeal, making them suitable candidates for evaluation via visual prototyping. On the other hand, the
remote controls that are used to operate such products can probably not be
effectively evaluated using visual prototypes. This is because users' responses
to such products are likely to be heavily dependent on the remotes' tactile
properties and interaction styles.

127

Models

Models are physical representations of a product. Typically, these will be made of wood or of polystyrene foam. Models can be particularly useful for assessing whether the proposed product would fit into its environment of use, as well as for checking whether its physical dimensions are suitable for the product's purpose.

In their most basic form models tend to be most appropriate for evaluation of the visual qualities of the product. However, more sophisticated models can be used as a vehicle for evaluation of the product's tactile elements. For example, models may include additional materials to add weight to the model so that its weight is accurately representative of the proposed product. The surface may also be finished such that it reflect how the product would feel to the touch. These issues can be particularly important when evaluating hand-held products, both in terms of the practical and hedonic aspects of how the product is experienced.

For example, the handset of a telephone must be of suitable dimensions and weight for a user to be able to hold it comfortably for a fairly protracted period. It is also important that the feeling of the receiver in the hand radiates a feeling of quality to the user. Similarly, the tactile properties of products such as kitchen utensils should radiate a sense of quality and should give the user a sense of confidence that he or she is in control. This might mean, for example, that knives should be well-balanced and heavy enough to 'help' the user slice through food. These principles have been central, for example, to the success of the range of kitchen tools made by Global Knives (see Figure 3.16).

Screen-based interactive prototypes

These are screen-based representations of products. They offer simulated interactions – for example, by positioning a cursor on a representation of one of the product's controls. The screen-based representation will then change state to represent how the product would react to a particular interaction.

This type of prototype is particularly useful when there are firm ideas for the form of the product and potential interaction styles, but sufficient uncertainty exists to make it worth checking the interface style before going on to build a full working version of the product itself. For example, de Vries and Johnson (1999) used screen-based interactive prototyping to evaluate the interface to a car stereo. They checked whether users could understand what each of the controls was for and whether they could work out how to use them. They also checked whether users found the information presented on the display meaningful. At this stage in the product creation process it may still be possible to make alterations comparatively cheaply, particularly

Figure 3.16 Global Knives kitchen tools

where the product has a software-based interface that could easily be repro-grammed.

Screen-based prototypes can also be an excellent medium for evaluating a number of different colour proposals for a product. Illustration programs such as PhotoShop support quick and easy re-colouring of an on-screen image, enabling the product-creation team to mock up a number of proposals for evaluation. For example, Kim and Moon (1998) used screen-based interactive prototypes in their investigation of people's emotional responses to colour use in the interfaces of cyber-banking systems.

Fully working prototypes

These are prototypes that – at least from the point of view of the user – are barely distinguishable from a fully manufactured product. In the case of software-based products, these will be the same as screen-based interactive prototypes – software that supports interaction and responds in the same way that the final product would do. However, where the proposed product has physical dimensions, the prototype will have the same dimensions and design elements – including form, colour, surface finishing, product graphics, etc. They do, then, enable people to experience the product fully, just as if it were the real thing.

An advantage of fully working prototypes is that they give people the chance to experience a product in its totality. This is in contrast to other sorts of prototypes that will only give a partial representation of the product – for example, the visual prototype will only give the visual aspects of the design, whilst a screen-based interactive prototype gives little or no indication of a product's tactile qualities. Experiencing a product in its totality can be important because a product may be experienced as more than a sum of its parts. Because of this it may not always be possible to predict the overall experience that people will have with a product, simply by summing up their reaction to the individual elements of that product. For example, whilst a person might like the form of a product and the interaction design, he or she might feel that in combination they are unsuitable. If the form and interaction design were to be tested separately – perhaps by evaluating a model to check the form and a screen-based interactive prototype to check the interaction design – then any incongruities arising from their combination may not be spotted. This is an argument for using fully working prototypes in evaluations. However, the problem with fully working prototypes – especially those for hardware-based products – is that they tend to be very expensive to produce.

In the case of a stereo system, for example, a fully working prototype would be one where the experience of use was no different to that of using a fully manufactured stereo. This would probably mean that the technological workings of the product would have to be similar to the final version of the product – a considerable development cost.

One way around having to implement technology fully at this stage is to develop 'Wizard of Oz' prototypes. These are prototypes that appear to be working in the same way as they would when manufactured, but in which the responses of the prototypes are actually being manipulated by the investigator or an associate (Jordan 1998). In the film, *The Wizard of Oz*, Dorothy and her friends (the Lion, the Scarecrow and the Tin Man) meet the 'Wizard' who appears to be a huge man with special powers and a deep, booming voice. However, it turns out that what they have really been confronted with is simply a machine behind which a man (the real Wizard) is standing, pulling levers. Similarly, with Wizard of Oz prototypes the users are unaware of the investigator's influence and think that the machine is responding to their inputs independently.

This type of prototyping tends to be most effective with software-based products. Consider, for example, an 'intelligent' information database such as an expert system. In an evaluation situation users might assume that they were interacting with a PC-based software package. In reality, however, they may simply be connected to another terminal and the responses to their inputs might actually have been typed in or selected by an associate of the investigator. However, as far as the users are concerned the responses are generated by the software and they can rate the system on this basis.

Some examples of the application of Wizard of Oz prototypes and of issues arising from their use can be found in papers by Maulsby *et al.* (1993), Vermeeren (1996) and Beagley (1996).

Evaluation criteria

When conducting an evaluation of a design it is important to be clear about the criteria against which the design is to be evaluated. A starting point for setting evaluation criteria is the product benefit specification. After all, the quality of a design should ultimately be judged on the basis of the degree to which it is able to deliver the benefits contained in the product benefits specification. For the purposes of evaluation, a series of 'goals' can be derived from the product benefits specification. These goals should represent specific, tangible criteria against which to evaluate.

Imagine, for example, that one of the items in the product benefits specification for, say, a video cassette recorder stated that the product should 'be quick and easy to install'. This might be translated into evaluation goals such as '90 per cent of people in the target group should be able to install the video cassette recorder within five minutes'. In order to determine whether a design concept met this goal, it would probably be necessary to conduct an empirical evaluation, using a fully working prototype of the video cassette recorder. If this was not feasible, for example for cost reasons, then the goal might be modified within the parameters of the sort of prototype that would be feasible. If, for example, time and financial constraints dictated that nothing more than a visual prototype could be afforded, then the goal may have to be modified to '90 per cent of people in the target group expect that, based on the appearance of the product, it would be quick and easy to install'.

Case study part 3: setting evaluation goals for a photo-camera

To further illustrate the setting of evaluation goals, return to the example of the photo-camera designed for use by young European women of high socio-economic status. As a reminder, the product benefits specification is listed again in Figure 3.17.

Camera should feel good in the hand
Camera should be easy to carry around
Camera fits well and comfortably against the face
Camera should be operable without causing damage to the users' fingernails
Camera should confer the impression of high socio-economic status on the user
Camera should confer the impression of high cultural status on the user
Camera should enable the user to take photos quickly
Camera should be operable without disturbing others or embarrassing user
Camera should be easy to use at the first attempt
Camera should give aesthetic pleasure
Camera should reflect the users' femininity
Camera should be environmentally 'safe'

Figure 3.17   Product benefits specification for a photo-camera designed for European women of a high socio-economic status aged between twenty-five and thirty-five years

First, consider the issue of the sort of prototype that should be used in this evaluation. In the second part of this case study, when the issue of how to define the properties of the product in order to deliver these benefits was considered, much emphasis was put on the aesthetic elements, such as the form, materials, colours and graphics. It seems, then, that for the purposes of evaluation, it may be important to have a prototype that embodies these aesthetic qualities. Although the properties of the interaction design were considered important in delivering some of the specified benefits, less emphasis was placed on this element than on the aesthetic elements. Given this, it may be acceptable to use a form of prototype that puts less emphasis on interaction design. In this case, then, it may be appropriate to make a model of the camera. Because tactile aspects of the camera were considered important, it would probably be advisable to ensure that the model was weighted and finished so as to accurately reflect these aspects.

Given that the sort of prototype used is to be a model, the evaluation goals by which design concepts are to be judged can be set accordingly. Suggestions as to evaluation goals are given in Table 3.8. Each evaluation goal corresponds to a specified benefit. In this case it is assumed that design concepts will be evaluated empirically using a sample of participants representative of those for whom the product is designed.

Before leaving this example, it is worth noting a couple of points about the evaluation goals suggested in Table 3.8. The first thing to note is that, in the context of the type of prototype being evaluated, some of the goals seem

Table 3.8 Suggested evaluation goals for a photo-camera designed for European
women of a high socio-economic status aged between twenty-five and
thirty-five years

Specified product benefit	*Evaluation goal*
Camera should feel good in the hand	80% of evaluation participants should indicate that they enjoy holding the model of the camera
Camera should be easy to carry around	80% of evaluation participants should indicate that they would expect the camera to be easy to carry around
Camera should fit well and comfortably against the face	80% of evaluation participants should indicate that it feels comfortable to hold the model against the face as if they were taking a photo
Camera should be operable without causing damage to the users' fingernails	80% of evaluation participants should indicate that they would not expect to damage their nails when operating the camera
Camera should confer the impression of high socio-economic status on the user	80% of evaluation participants should indicate that they feel that owning such a camera would confer the impression of high socio-economic status
Camera should confer the impression of high cultural status on the user	80% of evaluation participants should indicate that they feel that owning such a camera would confer the impression of high cultural status
Camera should enable the user to take photos quickly	80% of evaluation participants should indicate that they would expect that it would be quick and easy to take pictures with the camera
Camera should be operable without disturbing others or embarrassing user	80% of evaluation participants should indicate that they would not expect that operating this camera would cause embarrassment to them or disturbance to others
Camera should be easy to use at the first attempt	80% of evaluation participants should indicate that they would expect that the camera would be easy to use at the first attempt
Camera should give aesthetic pleasure	80% of evaluation participants should indicate that they find the design of the camera to be aesthetically pleasurable
Camera should reflect the users' femininity	80% of evaluation participants should indicate that they feel that the product has a positive, feminine image
Camera should be environmentally 'safe'	80% of evaluation participants should indicate that they feel that the camera would be environmentally safe

'better' than others in terms of their suitability as criteria by which to judge whether the proposed design delivers the specified benefits. In this case, for example, the goals associated with aesthetic and cultural benefits, such as the feel of the camera in the hand, the aesthetic appeal of the camera and the status conferred by the camera, might be expected to be accurate mirrors of whether the camera would, indeed, deliver these benefits. In contrast, goals associated with the usability of the camera and, in particular, the goal associated with whether the camera would cause embarrassment or disturbance seems to require the evaluation participants to make assumptions about a number of issues which would not be directly communicated by the model.

It is, perhaps, inevitable that any particular prototyping medium will be more suited to the evaluation of some aspects of a design and less suitable for the evaluation of others. In this case, the choice of a model over, for example, a screen-based interactive prototype favours the evaluation of the design's aesthetic aspects over its functional aspects. This is a trade-off, the like of which may be inevitable when selecting a prototyping medium. Depending on, for example, the budgetary constraints under which the evaluation is to be conducted, it may be possible to develop more than one prototype. For example, in this case, it might be possible to develop both a model of the camera and a screen-based interactive prototype. If an interactive prototype were to be used in addition to the model, then it may be possible to re-frame some of the goals relating to the functioning of the camera. For example, the goal associated with the benefit 'should be easy to use at the first attempt' might be re-framed as '80 per cent of evaluation participants should be able to work out how to load the film and take a picture within one minute of first seeing the prototype'. Similarly, if sound is included in the prototype, the goal associated with the benefit 'should be operable without disturbing others or embarrassing user' could be re-framed as '80 per cent of evaluation participants should indicate that they feel that the sound of the motors driving the camera would not cause embarrassment to the user or disturbance to others'.

Perhaps the only sort of prototype that would not require trade-offs of this nature is the fully working prototype. However, as indicated earlier, such prototypes can be extremely costly to make. This cost may, depending on the development budget available, be prohibitive.

The second point to consider is the clause '80 per cent of evaluation participants' at the beginning of each of the goals. Why has 80 per cent been chosen? Clearly, given the nature of some of the goals set, it seems unrealistic to expect that 100 per cent of respondents would have the same opinion about the design in relation to some of these goals. In particular, issues related to aesthetics will be largely a matter of opinion and taste. When deciding what percentage of respondents should agree in order to satisfy a goal, the evaluator should consider the implications of goals being met or not. In this case, the product is a consumer product – presumably the manufacturers want to satisfy as large a percentage of their target group as possible. Eighty per cent may seem a reasonable figure in this context – but ultimately it is a judgement call on the part of the evaluator, the manufacturer and those in the product development team.

An important point to note is that a photo-camera is not, in normal circumstances of use, a safety-critical product. As none of the goals is safety-critical, it may be acceptable that they are not met 100 per cent. However, where the product is safety critical, then it may be important that 100 per cent goals are set. For example, imagine that the product under development were a car stereo. The operation of such a product can distract people from driving and it is important that this distraction is minimised. An evaluation goal associated with such a product may be of the nature of 'evaluation participants should not take more than ten seconds to tune the radio to a particular channel'. In this case, it might be appropriate that, in order to fulfil the goal, 100 per cent of evaluation participants should be able to complete this task within ten seconds, rather than only 80 per cent. After all, a product that puts the lives of 20 per cent of its users at risk would be far from acceptable by any sensible standards of safety.

4

METHODS

In Chapter 3 the stages involved in creating pleasurable products were outlined. In this chapter, a number of methods for use in the product-creation process are described. The methods can be used for understanding people, for understanding what benefits they want from products, for understanding how to deliver this through the design and for evaluating design concepts. The methods described are not all equally suited to each of these purposes. Each of the methods has particular properties that affect its suitability for use at different stages of the design process and under different operational constraints. These include, for example, the time, effort and level of skill required to use the method, the facilities and equipment required and the number of participants needed in order to gather useful information. They also differ in terms of the sorts of information they are suited to gathering – for example, some may be particularly useful in understanding people, others in evaluating concepts, others in setting the product benefits specification and others in specifying the experiential or formal properties of a design.

The basic structure of each method is outlined, along with hints as to when it is most beneficial to use each of the methods and the extent to which a product concept must be developed before each method can be applied. The methods also differ in terms of the nature and quality of the data that can be gathered from their application. Some may, for example, be well-suited to gathering accurate noise-free data, whilst the strength of other methods may be in their utility in gathering data that reflects the real-life context in which the product is experienced. Similarly, some methods may be more suitable for gathering quantitative data and others qualitative data.

Two of the methods are non-empirical. These methods require no participants at all – the investigator simply uses a structured approach in order to make judgements about how pleasurable or displeasurable a product is. However, most of the methods do require participants, and these are known as empirical methods. In the context of usability evaluation, it has been argued that there is no substitute for seeing people using or trying to use a product (Jordan 1998). The same may be true of evaluating the pleasurability or otherwise of a product. Although taking a structured approach, such as that outlined in Chapter 3, should help in creating pleasurable products, there may be cases where people will react to products in ways that the 'expert' evaluator would find difficult to predict. An inherent advantage of empirical methods is that they can reveal such unexpected reactions.

Because pleasure-based approaches to human factors are comparatively new, the advantages of empirical methods may be particularly acute. This is because the body of knowledge relating pleasurableness to particular product properties is comparatively small when compared to, for example, the vast literature on the link between usability and product properties. Until such links have become more strongly established across a wider range of products and contexts, empirical evaluation of concepts may, in general, be the most sensible option. Nevertheless, there are circumstances in which empirical evaluation may be impractical or inappropriate. For example, if confidentiality is an issue or in circumstances in which it is very difficult to find appropriate evaluation participants.

Many of the methods described have previously been used for usability evaluation and are described here in an adapted form for measuring pleasure with products in a holistic sense. A number of the methods have their roots in psychology (for example, questionnaires, interviews, experience diaries), some have been adapted from marketing and market research (for example, focus groups, participative creation) and others have been developed specifically within the discipline of human factors (for example, co-discovery).

The aim of the rest of this chapter is not only to describe the methods, but also to give a flavour of what it is like to use the methods and advice about how they can be implemented most effectively. The empirical methods are presented first. These are not presented in any particular order. The non-empirical methods are then reported – again, these are not presented in any particular order.

Towards the end of the chapter, examples will be given of how some of these methods have been applied in a variety of approaches to the creation of pleasurable products.

Empirical methods

Private camera conversation

This method involves participants entering a private booth and talking to a video camera about a product or product concept. The investigator might give the participant a sheet containing a list of issues to talk about, or may allow the participant to decide for him- or herself what issues to cover. Typically, they will take the product or concept into the booth with them. Clearly there may be some situations in which this is not practical – for example, if the product under evaluation were an automobile or some other large product. If the concept has been developed to a level that allows the participant to interact with it – that is, if there is a model, screen-based interactive prototype, fully working prototype or finished product – then he or she would usually be given time to interact with the product ·beforehand. Perhaps the investigator will give the participant some specific tasks to perform or simply give him or her the chance to explore the product or concept at his or her own discretion.

The method can also be used at an earlier stage of the product creation process, for understanding people or for understanding the benefits required from a product and the properties associated with these benefits. For example, participants may be asked to talk about their favourite products and what it is that makes these products so special.

A variant on the method is to have two people in the booth speaking to the camera at the same time. This can have two potential advantages. First, the participants can prompt each other by picking up on points that the other has made. For example, if one participant were to mention a difficulty that had occurred when using a product then this might help the other participant to recall similar difficulties that he or she may have had. Similarly, if one participant gives an opinion about a particular aesthetic or functional feature of a product then the other participant may be triggered to comment on the same or a related aspect. Another advantage is that the participants might find it more easy to talk with another present, rather than simply talking to a video camera, where, of course, they will receive no direct response or feedback about what they are saying. A disadvantage, though, of having another present may be that there are interaction effects between the participants. This might lead to one participant dominating the session, whilst the other gets little opportunity to speak, or one participant effectively setting the agenda for discussion. It could also be that the presence of another will inhibit participants in terms of expressing themselves as freely as they might otherwise have done.

Advantages

Because the investigator is not present whilst the participant talks to the camera, this should minimise any potential investigator/participant interaction effects. This may mean that the participants will be less restrained in their comments than in a situation where they were speaking directly to the investigator. If the investigator meets the participants beforehand, then there may still be some effect from this; however, it should not be so great as it would be if the investigator and participant were face to face for the whole session.

This can be a particularly important issue when discussing issues connected to the more emotional and hedonic aspects of products. It has been noted, for example, that people will often feel more comfortable discussing only the rational reasons for reacting to a product in a particular way when in face-to-face discussions with an interviewer. The developers of the private camera conversation – Vries *et al.* (1996) – note that participants tend to talk far more widely and freely about such issues with this method than with face-to-face methods.

Many participants find the private camera conversation sessions enjoyable to take part in. The atmosphere here is, perhaps, a little less formal than with some other investigation methods and the idea of having the chance to be recorded on video can be appealing to some. This can be good for the public relations of the organisation conducting the evaluation. Vries *et al.* report a positive response on using the method at an exhibition centre and in a high school. Because people enjoy taking part it is also comparatively easy to recruit evaluation participants.

The video tapes themselves can make good 'evidence' when reporting back to a commissioner about the outcomes of the evaluation. Having no investigator present can also be helpful in this sense as there is little room for debate about whether the participant is being 'led' in any way.

The private camera conversation is a versatile method in that it can be used to evaluate prototypes at any stage of development – from a rough sketch or rendering of a concept, through to a finished product.

Disadvantages

The downside of not having any participant/investigator interaction during the session is that the investigator cannot control the direction in which the session goes. Thus, if the participant's monologue starts to go in a direction that is not relevant in the context of the issues under investigation, then there is nothing that the investigator can do to move it back on to the important issues.

Because there may be little structure to each participant's monologue, and certainly very little structure across the monologues as a whole, analysis of the sessions can be both complex and time consuming. Interpretation of certain of the participants' statements can also be difficult, as when analysing the tapes it is too late to question the participant as to what he or she meant when something seems ambiguous. For example, if the participant were to say something like 'I don't like the look of this product', then it might not be clear if he or she were referring to a dislike of the form, the colour, the product graphics, all three of these or some other aspect of the product's appearance.

Co-discovery

This method – described by Kemp and Gelderen (1996) – involves two participants working together to explore a product or concept. The idea is that by analysing the participants' verbalisations the investigator can gain an understanding of how the participants experience the product or concept. Usually the participants are friends of each other or at least acquaintances. This is beneficial, because if they know each other they are less likely to feel inhibited in speaking to each other about what they are doing and about their opinions of the product.

The investigator may sit with the participants when they are exploring the product, perhaps giving instructions, helping whilst they use the product or asking questions about what they are doing and thinking. Alternatively, the investigator may simply issue the participants with instructions beforehand and then retire to an observation room to monitor the session or, alternatively, record the session on video whilst he or she is absent. The instructions might be of a general nature, such as to explore the product under investigation, or they may request that participants complete particular tasks.

As an example, consider two people participating in a co-discovery session to investigate an audio system. The investigator might ask them to first explore the system and then to do some specific tasks, such as playing a cassette tape or finding a particular track on a compact disc. He or she might also ask the participants for their opinions about some aesthetic or functional aspect of the product. The co-discovery method is particularly suited to looking at people's initial responses to products. As well as observing which of the functional aspects of the product participants initially explore, the investigator can also hear what participants' first impressions of the product are.

Advantages

As with private camera conversations, the co-discovery method is comparatively informal in its set-up. When speaking with a friend or acquaintance participants may feel less pressure to 'rationalise' their views and comments than they would when speaking to an investigator.

Video recordings of co-discovery sessions can provide convincing material to show to those using the results of an investigation. Because the conversation is between participants and thus perhaps more spontaneous than if the participant were talking to the investigator, this may convince those interested in the outcomes of the investigation that what is being said is unprompted and represents the real concerns of the participants. For example, if the participants' discussion centres mainly on the aesthetic qualities of a product, then it might be concluded that, for these people and in the context of this product type, it is the aesthetic qualities rather than the functional qualities that are uppermost in influencing the overall pleasurability or otherwise of the product.

Disadvantages

Because this method is one with which the participants will have a large degree of control over the topics covered in the discussions between them, it may not always be possible for the investigator to control the direction in which discussions go. There is, then, no guarantee that all of the issues which the investigator wishes to cover will be raised. Clearly, there is a trade-off that can be made here. If the investigator wishes to be certain that particular issues are raised, he or she could sit with the participants during the evaluation session and ask about these. Alternatively, he or she could include a request to cover these in the instructions initially given to the participants. However, the more influence the investigator has over how the session proceeds, the less spontaneous it is likely to become – and one of the major potential advantages of the technique may be compromised.

Focus groups

The focus group – a technique originally developed within the discipline of market research – is a group of people gathered together to discuss a particular issue. The discussions could cover, for example, users' experiences of using a particular product, their attitudes towards the aesthetics and functional aspects of a particular product, information about the contexts in which they will experience the product and the types of pleasure or displeasure that they associate with a particular product, or simply their general preferences and dislikes with respect to product design.

A focus group consists of a discussion leader and a number of participants. The leader will have an agenda of issues that will form the borders within which the discussion can proceed. This agenda is usually rather loosely structured as the aim is to allow participants to take the lead in determining the direction in which they wish the discussion to go. This should ensure that the points raised will be those that are of most concern to the participants. In facilitating the focus group, the leader's job is to ensure that all participants have a chance to voice their opinions. It may be, for example, that some of the participants are more vocal than others, and it is important to prevent one or two people dominating the discussions to the exclusion of the others.

It is also usual for the leader to have a set of prompts. These are for use in the event of the discussion 'drying up' due to participants not being able to think of anything useful to say. It is, however, important that the prompts are simply means of triggering more conversation and that they do not lead the participants into giving particular responses. Making a prompt effective can come down to subtleties in the language used. For example, if discussing, say, the surface texture of a hand-held product such as a remote control, it would be inappropriate to prompt with 'Don't you think the tactile properties of the remote give it a high quality feel?' Rather, it would be more appropriate to use a prompt such as 'When you hold the remote does it feel of a low or high quality and why?' The first prompt is loaded, as the leader has phrased it in such a way that it may give the impression that he or she believes that when held the remote gives the impression of high quality due to its surface finish and is asking the participants to agree. The second prompt, however, is phrased in a neutral way. The leader's phrasing doesn't give the impression that he or she is expecting a particular answer, but comes across as a genuine query. Such a prompt simply gives participants something concrete to discuss and should serve to restart the conversation. Prompts should, however, be used only when there seems to be a problem in continuing the conversation and not as a means of redirecting a conversation that is in full flow. Even neutrally phrased prompts have the drawback that they can lead participants into discussing issues that might not have been particularly important to them.

As with all techniques involving open-ended questioning, a problem in analysing the discussion in a focus group is in interpreting why a particular issue has not been mentioned. In the example just given, the reason why the tactile properties of the remote had not been discussed by the participants until prompted might be that the issue was of no real interest to them, or it might simply be that nobody in the group had thought of the issue until prompted. The enthusiasm with which the conversation proceeds directly after the prompt can be an indicator of this, but if the leader is in any doubt, then the best thing to do is simply to ask how important the issue is after the prompted part of the discussion is over.

When deciding on the number of participants to include in a focus group, the investigator has to consider the trade-off between two main factors. The more people who participate in the group, the more chance for participant interaction. Indeed, one of the main advantages of focus groups is that one participant's comments can trigger a useful contribution from another participant. Clearly, the more people that participate in the group, the greater the chance of this happening – if there are too few participants then this effect may not be achieved.

The other factor, though, is that of giving all participants a chance to voice their opinions. With respect to this, it is beneficial to have fewer participants, because if people are having to wait for too long before getting a chance to talk they may get bored or frustrated. This may lead to them feeling excluded from the proceedings and make them unwilling to contribute. It is difficult, and probably unwise, to make a general statement as to the 'right' number of participants to have. Typically, when used in market research, focus groups have tended to include between eight and twelve participants (Jordan 1998). However, the tradition within human factors – where focus groups have been used to investigate usability issues –

is to include less participants – typically five or six (see, for example, O'Donnell *et al.* 1991 or Jordan 1994).

As with any empirical investigation method, care must be taken in selecting those who are going to participate in the focus group. Because of the interpersonal dynamics that are involved with this method, it may be tempting to select participants who might be expected to be particularly vocal in expressing their opinions. This, however, puts a bias on the sample and should not be necessary. Rather, it is the job of the leader to make sure that all participants get involved, no matter how reserved they might be. Learning how to be an effective focus group leader is, perhaps, as much of an art as a science and is most effectively learnt through experience and through watching others.

Advantages

Focus groups can be used at any stage of the design process – participants can discuss a concept, a visual or working prototype, or experiences of using finished products. Because the method is loosely structured, participants have the opportunity to raise issues that the investigator may not have anticipated to be important. The group dynamics involved can be particularly beneficial here as an issue raised by one user may stimulate ideas from others

Disadvantages

Whilst there are potential benefits due to the group dynamics, these may also bring disadvantages. For example, there is a danger that one or two of the members of the group may prove particularly dominant. This may mean that the opinions that are apparently those of the group as a whole may, in fact, simply reflect the opinions of this individual or individuals. Similarly, there may be someone in the group who is particularly retiring. This person may stay quiet during the discussions and, thus, may not get his or her opinions heard. Both of these problems can be addressed by the leader of the focus group. For example, the leader can make a particular point of asking quieter group participants to give their opinions by directly addressing questions to them. Similarly, the leader can ensure that particular individuals do not dominate the conversation by politely acknowledging their contributions and then addressing a question to another group member. Managing the group dynamics is one of the skills that is necessary in a successful focus group leader.

Think aloud protocols

With this method participants are asked to verbalise the thoughts that they have when experiencing a product or product concept. Where the item under evaluation is a finished product or an interactive prototype, participants may be set some specific tasks to do or may be given the chance for some free exploration of the product.

During the think aloud session the investigator will usually prompt the participant, in order to encourage him or her to make helpful verbalisations. These prompts may simply be of the general type, for example, 'What are you thinking now?', or they may be more specific, perhaps relating to a particular aspect of the product's aesthetics, functionality or interaction style. An example of a more specific prompt might be, 'What was your initial reaction when you saw the colour of the product?' or 'How do you feel the graphics affect the overall appearance of quality?' Participants' verbalisations may also give information about how they respond emotionally to a product. For example, a prompt might be, 'How would it feel to own this product yourself?'

Advantages

From participants' verbalisations it is possible to understand not only how people react to a product, but also *why* they react in this way. This means that they can be an excellent source of prescriptive data, which can lead directly to design solutions. For example, when redesigning a high-end telephone, Jordan and Kerr (1999) used think aloud protocols as a method for investigating the suitability of functionality and interaction style in a prototype telephone system. Here, participants' verbalisations were useful indicators of the way in which users might see the relationship between functions and thus the way in which functions could most effectively be grouped.

Think aloud protocols can also be an efficient way of getting a lot of information from only a few participants. Virzi (1992) cites the think aloud protocol as one of the most efficient human factors methodologies in terms of information gathered per participant. This is because each participant can provide such rich prescriptive information.

Disadvantages

A potential disadvantage is that because participants are explaining their views or actions to an investigator, they may feel tempted to 'rationalise' as to why they are responding to a product in a particular way. They may also feel that they want to come across to the investigator as a 'nice' person and may thus avoid answers that they feel may show them in a bad light. For example, if a participant were asked why they liked the design of, say, a television set, and the real reason was that he or she felt that it was ostentatious and would be the envy of their neighbours, then he or she may well be tempted not to say so! Perhaps, instead, he or she might say that it would blend in well in the living room or that the functions were particularly appropriate. When analysing the outcomes of a think aloud protocol it is important to be aware of such potential effects that could be operating on participants' verbalisations. The danger otherwise is that a false picture may emerge about the sorts of things that are important to people with respect to a particular product. In particular, the importance of a product's 'rational' qualities may become exaggerated at the expense of the emotional ones.

As with focus groups, there is a danger inherent in getting the level of prompting wrong. Whilst too little prompting might result in missed information, too much prompting might encourage participants to comment on issues that were really of very little significance to them. Similarly, they may make up reasons for liking or disliking a particular aspect of a product, when, in reality, they may not really be sure why they like or dislike this aspect. For example, imagine that the product concept under evaluation were a microwave oven, and the participant happened to like the style of the graphics on the product. If asked why, then he or she may be tempted to offer a reason such as 'they look modern', or 'they go well with the form of the product'. In reality, however, they may just happen to like them without any 'real' reason. Again, there is a danger that false connections are made between experiential and formal product properties and incorrect conclusions may be drawn unless the investigator is aware of this potential pitfall. Having a feel for what is the right level of prompting is, then, a skill that is central to facilitating an effective think aloud session.

Experience diaries

Experience diaries are mini-questionnaires that are issued to users in order that they can make a note of their experiences with a product over a period of time. Participants may be asked, for example, to fill in a page of the diary each week to record, say, how they feel about the product, what they particularly like or dislike about the product, any problems they may have when interacting with the product and what their general impressions of the product are.

In most cases, the diaries will be issued to people to fill in outside the presence of an investigator. Because of this it is important to ensure that each diary entry will not take too long to complete. If people see completing an entry as a chore, then they may soon stop doing it, especially if there is no investigator present to encourage them. There is, of course, a trade-off here. Whilst it is important to keep each entry short, it is also important that the user records enough information to make the diaries useful. This means that when designing experience diaries it is vital to have a good idea about the comparative importance of the various sorts of information that could be gathered, so that the vital questions can be included in the diary and the less vital ones left out.

Which questions are vital might depend on, for example, the level of confidence within the product development team with respect to decisions made about the product's design. If, for example, the team were confident about the aesthetics and functionality, but not about the interaction style chosen, then the diary might mainly contain questions relating to usability issues. On the other hand, if the product was one whose appeal may have a significant socio-pleasure component – such as a mobile phone – then the diary might contain a number of questions about the reaction of others to the product.

Experience diaries can be very useful for capturing relatively infrequent events. For example, if a laptop computer were the product under evaluation, then the diary might ask about a person's experience with the laptop during different types of journey – for example, a rail journey, a boat journey, a car journey and an air journey.

Advantages

The method is cheap in terms of investigator time and effort, as having decided on a set of questions, the diaries can then be sent to as many people as necessary. They are also cheap in terms of the facilities needed to administer them. No laboratory, video or audio facilities are required.

Another major advantage of the method is that it can be used to monitor how people's experiences with a product change over time. Most evaluation methods tend to give a 'snapshot' of a person's experience with a product – perhaps a first impression or an impression as it stands after a particular period with a product. By giving information about how things change, data gathered from an experience diary can also help the investigator to deduce how the effects of particular aspects of a product's design will affect people differently at different stages of their experience of the product.

Consider, for example, a portable stereo system – a 'boombox'. In the initial stages of ownership of such a product, a person's first reaction may be one of, for example, excitement and pride. It may also be that the elements of the product that lead to this reaction are, say, the performance and aesthetic elements. So, then, the good sound quality and the interesting form, colours and graphics may lead to this sense of pride and excitement in ownership.

As the product gets older, the person's relationship with the product may change. Instead of pride and excitement, perhaps, if the person–product relationship remains a good one, the person will feel, for example, a sense of attachment to and reassurance from the product. This might be connected to, for example, the functionality, reliability and usability of the product. If, over time, the person has found that the stereo has just the functions that he or she wants, that these functions are easy to use and that the stereo will always perform reliably, then he or she is likely to become attached to the product and feel reassured that it will 'do its job'.

In such a case, then, the qualities that attracted the person to the product in the first place are different from those that will cement a good person–product relationship in the long term. It is very important for manufacturers to be able to understand this distinction, as a person's initial reactions to a product are likely to determine whether he or she buys the product in the first place. Meanwhile, the longer-term reactions are likely to have a major influence on whether the person will buy another similar product or another product from the same manufacturer.

Disadvantages

Despite the best efforts of an investigator when designing an experience diary, there is still no guarantee that people will complete them at the pre-prescribed time or incident points. Even if this could be guaranteed, there is no guarantee that they will be completed accurately. Sometimes the vocabulary that people use to describe experiences may be difficult for the investigator to interpret. For example, returning to the example of the portable stereo, if the participant were to make an entry such as 'I like the stereo because it looks so good', then it may be rather ambiguous as to what conclusions the investigator should infer from such a statement.

In the first instance, it is difficult to deduce what the person means by 'liking' the stereo. Does this mean that they are proud to own the stereo, that they feel excited by the stereo or that they feel comfortable or reassured by the stereo? Similarly, what is meant by 'it looks so good'? Is this a reference to the form, the colours, the graphics, the materials, the dimensions or all or none of these?

This problem is not unique to experience diaries. Indeed, any method that requires participants to give information in the absence of an investigator is likely to be open to the same problem. One way of addressing such a problem is to provide the participant with a number of possible answers from which they can select the most appropriate response. However, such an approach requires the investigator to anticipate the nature and range of responses that the participant may have and the sorts of product properties and elements that might engender such responses. This issue will be discussed at more length later in this chapter, particularly in the section on questionnaires.

Another disadvantage of this method is that it can only be used effectively with finished products. Participants are noting their responses to finished products that they are using over a period of time. It is not likely to be meaningful to use the method with a product prototype. So, whilst the method can provide useful information for future generations of products, it is unlikely to provide information that can be used to improve the design of the product under evaluation. The exception to this rule might be in the evaluation of software-based products. Here it may be possible to issue fully working prototypes to a group of people and incorporate what is learned from their responses into the design of the final product.

Reaction checklists

In its most basic form, a reaction checklist is a list of potential reactions that a person may have to a product. The participants are simply asked to mark against the reactions that they have had or that they anticipate that they would have to a product or product concept. The list of potential reactions could include those in all four pleasure categories, depending on the type of product. For example, Figure 4.1 shows a fragment of a checklist that might be applied to the evaluation of a mobile phone; participants would be asked to check the items to indicate which reactions they have.

The fragment shown in Figure 4.1 shows a reaction checklist in its simplest form – simply a list from which participants can indicate the reactions that

---

Physiological pleasure

The phone feels good in the hand

The buttons feel good to the touch

The phone can be comfortably carried

Psychological pleasure

The phone has useful functions

The phone is easy to use

The phone is fun to use

Sociological pleasure

I feel proud when others see me with the phone

Owning and using the phone enhances my social image

I enjoy being permanently contactable via a mobile phone

Ideological pleasure

Having this phone makes me feel better about myself

I find this phone to be aesthetically pleasing

Figure 4.1 Fragment of a reaction checklist for a mobile phone

they have had when experiencing the product. Note the fragment shown included only positive reactions on the list. It can be equally useful to include a number of negative reactions on the list to elicit information about potentially displeasurable responses. Extended reaction checklists could ask for additional information, for example about the regularity with which a particular reaction is experienced or the times when, and the situations in which, the participant has a particular reaction to a product.

Checklists have a history of use within human factors, usually in the context of assessing which features of a product people most commonly use. Edgerton and Draper (1993) found that checklists offered considerable advantages over open recall in the context of asking respondents about the features which they used on computer software packages.

Another possible extension to the reaction checklist is to combine the checklist with a list of product properties – including functional, aesthetic and interaction style features. The participant is then asked to check the aspects of a product that he or she particularly likes or dislikes. It is then possible to make links between reactions and product properties. This could be done, for example, via a statistical technique such as a cluster analysis. Typically, these product properties are listed as written descriptions. In the case of a mobile phone, for example, properties listed might include: 'the colour of the phone', 'the material from which the buttons are made' and 'the weight of the phone'. An alternative is to provide visual representations of the product properties. This means giving the participant a photograph or drawing of the product, perhaps supplemented by written labelling, and asking them to mark against the properties that they find particularly pleasurable or displeasurable. In the context checklists used for investigation of feature use, Edgerton (1996) found that visual checklists offered advantages over written checklists in terms of the validity of participants' responses.

Advantages

Reaction checklists are a comparatively cheap evaluation method, both because they are undemanding of investigators' time and because they require few facilities – no laboratory or video equipment is necessary, for example. They provide an effective means of gaining an overview of a person's response to a product.

Disadvantages

The method does not provide data that gives a direct understanding of why a person reacts to a product in a particular way. Even with extended checklists, which enable links to be made between reactions and product properties, the data still does not directly reveal why these links are there. For example, imagine that analysis of a number of reaction checklists used in the evaluation of a mobile phone revealed that there was a link between pride in ownership and the colour of the phone – imagine, for the sake of the example, that this was blue. The investigator would be able to deduce that the blue colour of the phone contributed to people's pride of ownership, but he or she would not know why. Of course, knowing that there was a link may be extremely useful in itself. The investigator could feel confident that manufacturing this phone in blue would be likely to contribute to pride in ownership of this phone. However, it might be difficult to predict whether this finding was generalisable to other phones and other products.

If, for example, the participants enjoyed the blue colour because they felt that it went well with the form of the phone, then this result may not be generalisable to a different phone. On the other hand, if the blue colour was liked because of more general associations with the colour – for example, perhaps people feel that blue is a 'classy' or 'sophisticated' colour – then a more robust, more independent link between pride in ownership and blue colouring might be expected.

This disadvantage is somewhat compounded by a second disadvantage of reaction checklists – they are less effective when not being used with finished products, as it is usually important that the participant gains experience of a product over a period of time and in 'real-life' contexts before completing the checklist. This means that the information gained from a reaction checklist will usually be used in the design of a future product, rather than in the redesign of the product under evaluation. This, then, makes the issue of generalisability of results to another product an important one, requiring the judgement of the investigator.

Field observations

Field observation involves watching people in the environment in which they would normally experience a product. Because the investigator directly observes the participant, this gives a degree of validity that might be lacking in investigations that rely on participants' reports. The context of the study – the real context in which the product is experienced – can give a greater degree of ecological validity than investigations conducted in the somewhat sterile environment of a laboratory.

Sometimes the investigator will not give any instructions, but will simply let the participants get on with what they would do anyway. Sometimes, however, the investigator might give instructions of a somewhat general nature. For example, the investigator might ask participants to perform some particular set of tasks using the product. There are, though, few controls and balances involved when conducting field observations. The idea is to gain an understanding of how the product is experienced under natural conditions without imposing boundary constraints, which would arise with a set investigation protocol.

It is important, when conducting a field evaluation, that the investigator tries to ensure that the effect of his or her presence is minimal. If the participants are aware that they are being watched, they may consciously or subconsciously alter their usual reactions to the product. This would compromise the level of ecological validity. Perhaps the most effective way of minimising investigator presence is simply not to let the participant know that he or she is being watched. This could be done, for example, by viewing the participants from a distance, or by filming them with a hidden camera. However, taking such an approach raises ethical questions. Under the ethical standards commonly accepted by those conducting human-factors or psychological investigations, the right of participants to be informed as to what is going on is regarded as central. Investigations would not normally proceed without the participants' prior permission. One solution to this that might be acceptable in some circumstances would be to inform participants afterwards that data or video recordings have been taken and then ask for their permission to use these for analysis purposes.

Analysing data from field studies can be comparatively complex. Before actually observing users it can be difficult to anticipate what issues will arise and thus it may be difficult to decide a priori on measures by which product pleasurability can be judged. Similarly, the real context of use of a product can be such that measures of pleasure or displeasure, which would have given meaningful data in the context of a controlled laboratory environment, can prove too insensitive in these situations. This is because there may be many other factors that will affect the reactions of a participant at any moment in addition to the effect from his or her experience of the product.

Imagine, for example, that the product under investigation were a computer software package and that the context of use were an office environment. The investigator might observe, for example, the frequency with which the participant used the package, the sorts of tasks that he or she used it for, the difficulties that the participant has when using the package and the apparent mood of the person when using it. Each of these measures can potentially give an insight into how pleasurable or displeasurable the product is. However, in real-life situations, each of these may also be affected by factors that are separate, if not independent, from the pleasurability or displeasurability of the product.

For example, the frequency of usage may depend on how busy the person is that day. The sorts of tasks for which he or she uses the product may depend on the nature of the person's job. The number of difficulties in use may be associated with the number of distractions that there are in the workplace. The mood of the person may be affected by many other job-related or personal factors. It may be, then, that the effects of the product's pleasurability or displeasurability is swamped by a number of other factors.

It might be argued that if an effect is not large enough to show up in these circumstances, then it may not be worth bothering about. After all, if the effects of the product's pleasurability or displeasurability are outweighed by other factors, then do these effects really matter?

Unfortunately, it is often difficult to know the answer to this question. In the first place, it is not always easy to know whether a person's reaction is product related or not. Returning to the example of the software package, perhaps the reason that the person uses this package a lot is, after all, because he or she finds it pleasurable. Perhaps even when very busy, there are alternative ways of approaching the task that would not have required the use of the package. Perhaps his or her mood at the time was, indeed, a reaction to the software package.

Even if the investigator could be sure that the effect of the product itself on the person's reactions were only marginal, would this necessarily mean that it were unimportant? This is something that it may only be possible to judge over time. Consider, for example, a multi-function CD player. Imagine that the player has a number of sophisticated functions, for example an option to program the player to play tracks in a particular order. If people struggle with this once or twice the first few times that they use it, then they may bear this cheerfully. However, if they are still having problems with this on the tenth or eleventh time of using it, then they might find it extremely annoying and frustrating. Had the investigator observed the participant at an early stage with the product, then he or she might conclude that this problem were of minor significance to the person. However, observation at a later stage may have revealed a severe reaction.

Equally, an issue that appears trivial in one context of use may become sharply magnified in another context. Staying with the example of the CD player, problems in programming that may be accepted when a person is alone or with close family may become a source of acute embarrassment when trying to use the product in front of guests.

In the end. probably the most effective way of dealing with such ambiguities is simply to ask participants about them. However, the investigator is then returned to a situation in which he or she is relying on user reporting, arguably negating one of the main strengths of the field observation method. The art of running an effective field observation probably depends on getting the balance right between observation and asking. The investigator must balance these two to come to an overall conclusion about what is really going on.

Advantages

The central advantage of the field observation is that, despite the caveats mentioned above, this method is probably the one that comes closest to being an analysis of a product's pleasurability under 'natural' circumstances.

Disadvantages

The possible ethical difficulties are disadvantages, as are the difficulties in understanding how the effects from the product are balanced against other effects arising from the environment and context of use. Another disadvantage of field observations is that they are usually only carried out on finished products. In this sense they lack the flexibility of, say, questionnaires and interviews, which can be used throughout the design process. It would not usually be meaningful, for example, to consider using a field study to test a concept or an early prototype. However, there have been cases where field observation of performance with interactive 'Wizard of Oz' prototypes has proved beneficial (see, for example, Beagley 1996).

Questionnaires

These are printed lists of questions. Broadly speaking, there are two categories of questionnaire – fixed-response questionnaires and open-ended questionnaires. With fixed-response questionnaires, people are either presented with a number of alternative responses to a question and asked to mark the one that they feel is most appropriate, or they are asked to register the strength with which they hold an opinion on a scale. Consider the following example of a questionnaire item with which users are asked to choose from a selection of responses. In the context of asking about, for example, the easiness or difficulty of interacting with a product, a questionnaire might contain the statement 'this product is easy to use'. Respondents might then be asked to mark a box to indicate their level of agreement or disagreement with this statement. These boxes could be labelled 'strongly agree', 'agree', 'not sure', 'disagree' and 'strongly disagree'. Similarly, if asked how often they use a product, respondents might be given the choice of 'very often', 'quite often', 'occasionally', 'rarely' or 'never'. With this type of fixed response questionnaire it is important that the response choices given cover the full range of possible responses and that the wording can be understood by the respondents. It would, for example, be inappropriate if the alternative responses for a question about amount of usage were to be only 'very often', 'rarely' or 'never', as this would not give people who regarded themselves as 'occasional' users the opportunity to choose a category that they felt that they could agree with. Of course, unnecessarily complex language should be avoided. For example, if asking about how easy something was to use, then the word 'easy' would be more understandable than, say, 'elementary'.

Using numerical scales is one way of simplifying the task of using the appropriate semantics, as here it is usual to only use two semantic 'anchors' – one at each end of the scale. For example, the 'Task Load Index' (TLX) mental workload questionnaire developed by the North American Space

Agency (NASA) (Hart and Staveland 1988) is one that has a history of use within human factors, where it has been used as a means of assessing the mental demands associated with product use (see, for example, Jordan and Johnson 1993).

The 'System Usability Scale' (SUS) (Brooke 1996) is another example of a questionnaire that employs the technique of asking users to mark scales between two semantic anchors. Here, however, there are five distinct scale points to choose from. The questionnaire, which was originally developed for use in the context of computer systems, but which has been widely used to measure the usability of consumer products, lists a series of statements that the respondent then has the chance to agree or disagree with. These include, for example, 'I felt very confident using this system.' The scales are anchored with 'strongly disagree' and 'strongly agree'.

When designing fixed-response questionnaires to collect quantitative data, it is important to pay attention to the issues of 'reliability' and 'validity'. These are complex concepts, both in terms of how they are defined and how they can be measured. In broad terms, reliability is about the repeatability of what the questionnaire measures, whereas validity is concerned with whether the questionnaire measures what it is supposed to measure. In the context of using a questionnaire for evaluating the pleasurability of a product, reliability would relate to whether a particular respondent would give the same responses if asked to fill in the same questionnaire, in relation to the same product, on two separate occasions. If this were not the case then responses might be more of a reflection on, say, respondents' moods at the time of completing the questionnaire than on the pleasurability of the product that they had been asked to rate. Even given that a questionnaire is reliable, this does not necessarily mean that it will be a valid measure of pleasurability. A 'pleasure' questionnaire will be valid only if the questions and the responses given really probe the issue of product pleasurability. If the questionnaire is poorly designed it could be that responses will reflect some other aspect of the product such as, for example, its perceived monetary value, or its usefulness.

Using pre-prepared questionnaires, such as the TLX and the SUS, saves the investigator from having to be concerned with reliability and validity, as those who designed the questionnaires have already carried out checks on these. There are a number of pre-prepared questionnaires for measuring usability – SUMI (Software Usability Measurement Inventory) (Kirakowski 1996) and CUSI (Computer User Satisfaction Inventory) (Kirakowski and Corbett 1988) are excellent examples. However, as has been discussed throughout this book, whilst usability is likely to be an important component of what makes a product pleasurable, it is only one aspect of this.

The questionnaire given in Figure 4.2, on pp. 156–7 was designed to quantify product pleasurability (the total pleasure score can be calculated simply by adding the scores from responses to each individual question). It has been pre-validated and checked for reliability and has a history of use in the evaluation of consumer electronics products; however, it has never previously been published. Indeed, to date, the human-factors literature contains no other examples of pre-prepared questionnaires for the measurement of product pleasurability.

Whilst pre-prepared questionnaires such as the one illustrated in Figure 4.2 may give a good measure of a product's overall pleasurability, there may often be situations where the investigator wishes to design a questionnaire to address specific design issues. Then, of course, he or she may have to tackle reliability and validity issues.

With open-response questionnaires, respondents are asked to write their own answers to questions. For example, a question might be 'What are the best aspects of this product?' or 'What do you think of the appearance of this product?' Open-ended questionnaires can be particularly useful in situations where the investigator does not know what the important issues are likely to be with respect to a design's pleasurability. With fixed-response questionnaires the questions have to be framed specifically enough to make the response categories meaningful. With open-ended questions, however, questions can be framed more broadly, enabling the respondents to highlight the issues that they find most relevant.

Generally, open-ended questionnaires are, perhaps, more suitable for the early stages of a design, before the important pleasurability issues have been clearly identified. Indeed, the qualitative data that they can provide can play an important part in identifying these issues. In contrast, the quantitative data that can be obtained via fixed-response questionnaires can provide a metric by which to judge product pleasurability and, thus, fixed-response questionnaires are more commonly used after users have had a chance to use a new product, or at least an interactive prototype.

Advantages

An advantage of questionnaires is that having once designed a questionnaire and checked it for validity and reliability, the questionnaire can then be copied and issued to as many people as the investigator feels appropriate at little extra cost. They can, then, prove a cheap and effective method for gathering data from a large population. The method is also versatile in that it can be used at any stage of the design process. Questions can be formulated for investigating people characteristics or people's attitudes to prototypes or finished products. Because the investigator need not be present whilst respondents are filling in the questionnaires, questionnaires can also give the advantage of being free of investigator effects. With an interview, for example, respondents might consciously or subconsciously gear their responses to what they think the investigator wants to hear. The possibility for anonymity afforded by questionnaires can reduce or eliminate these effects.

PLEASURE WITH PRODUCTS (GENERAL INDEX)

1. I feel stimulated when using this product

0	1	2	3	4
Strongly disagree		Neutral		Strongly agree

2. I feel entertained when using this product

0	1	2	3	4
Strongly disagree		Neutral		Strongly agree

3. I feel attached to this product

0	1	2	3	4
Strongly disagree		Neutral		Strongly agree

4. Having this product gives me a sense of freedom

0	1	2	3	4
Strongly disagree		Neutral		Strongly agree

5. I feel excited when using this product

0	1	2	3	4
Strongly disagree		Neutral		Strongly agree

6. This product gives me satisfaction

0	1	2	3	4
Strongly disagree		Neutral		Strongly agree

7. I can rely on this product

0	1	2	3	4
Strongly disagree		Neutral		Strongly agree

8. I would miss this product if I no longer had it

0	1	2	3	4
Strongly disagree		Neutral		Strongly agree

9. I have confidence in this product

0	1	2	3	4
Strongly disagree		Neutral		Strongly agree

10. I am proud of this product

0	1	2	3	4
Strongly disagree		Neutral		Strongly agree

11. I enjoy having this product

0	1	2	3	4
Strongly disagree		Neutral		Strongly agree

12. Using this product helps me feel relaxed

0	1	2	3	4
Strongly disagree		Neutral		Strongly agree

13. This product makes me feel enthusiastic

0	1	2	3	4
Strongly disagree		Neutral		Strongly agree

14. I feel that I should look after this product

0	1	2	3	4
Strongly disagree		Neutral		Strongly agree

Overall Pleasure Rating

Figure 4.2   Pre-prepared questionnaire for quantification of product pleasurability

Source: Questionnaire designed by Patrick Jordan, Philips Corporate Design (1996).

157

Disadvantages

Possibly the biggest disadvantage of questionnaires filled in remotely from the presence of the investigator is that only a small proportion of them are returned and completed. The return rate for mailed out questionnaires is around 25 per cent (Jordan 1993). The reason why this is a problem is not the low number of questionnaires completed *per se*. After all, if the investigator wanted a sample of 100 respondents, then he or she could simply mail out 400 questionnaires. Rather, the problem lies in the likelihood that the people who actually take the time and effort to complete the questionnaire will not be a representative sample of the people in whom the investigator is interested. Those who complete the questionnaire will often be those with comparatively extreme opinions about the issues that are being asked about.

Consider, for example, a scenario whereby the manufacturer of software for use with home computers decided to survey its customers to check their levels of satisfaction with their products. It seems likely that those who would take the trouble to respond would be most likely to be those who had a particularly strong opinion about the software. If the manufacturers were to make the mistake of treating the questionnaires they received back as being representative of the opinions of their user population, they may come to the conclusion that their customers were firmly divided into two camps – those who loved the software and those who hated it. In reality, of course, these respondents are likely to simply represent the extremes of a much broader spectrum.

The problem of low response rates is likely to be exacerbated if the questionnaires issued are particularly long, as this will increase the time and effort needed to respond. Questionnaires that are to be completed remotely should, then, be as short and concise as possible.

One way to solve the problem of low response rates is to invite respondents to complete questionnaires in the presence of the investigator. Clearly, however, this would put demands on the investigator's time, thus negating one of the main benefits of questionnaire use. Another potential disadvantage of the questionnaire is that more care may have to be taken in the formulation of questions than in, for example, an interview. This is because, if the questionnaires are to be completed remotely, respondents will not have the opportunity to ask the investigator about anything that is unclear to them. In an interview situation, if there is any ambiguity in the wording of a question or in the meaning of the various response categories, then the interviewee can ask for clarification. With remotely completed questionnaires, however, the respondent must make his or her own interpretation of what is meant. Of course, if the questions are not clear, there is a significant possibility that misinterpretations will occur.

Interviews

Here the investigator compiles a series of questions that are then posed directly to participants, usually in a face-to-face situation or perhaps over the telephone. There are three broad categories of interview – unstructured, semi-structured and structured.

In an unstructured interview the investigator will ask respondents a series of open-ended questions. This gives the respondents the opportunity to steer the discussion towards the issues that they regard as important, rather than rigidly sticking to an agenda set a priori by the investigator. This type of interview may be most appropriate in situations where the investigator has little idea, in advance, of what the issues of concern to the people experiencing the product might be. Suppose that, for example, a high-end television set containing many new features – for example, to do with sound and picture control – was being considered for the market and that the manufacturers wished to gain a feel for which features were likely to bring the greatest benefits to the users and which were spurious. It might be appropriate, in this case, to simply ask participants who had a chance to interact with a prototype general questions such as what their favourite features were, which features they didn't like and why they liked or disliked a feature. There would, then, be little constraint on the sorts of reply that respondents could give.

With a semi-structured interview, the investigator would normally have a clearer idea of what he or she considered to be the relevant issues for an evaluation and thus of the sorts of issues that he or she might expect respondents to cover when answering questions. Respondents, then, would be a little more constrained as the investigator would try to ensure that certain points were covered by their answers. This is often done by prompting the respondents as they give their answers. Consider again the example of the high-end television. If the investigator was interested in users' responses to the features in general, but especially their response to some features in particular, then a general question could be supplemented with prompts. So, if the manufacturers were particularly interested in, say, how people would respond to a feature such as a sound equaliser, then the investigator might specifically ask about this feature as part of questions about what they thought the most and least pleasurable aspects of the product were.

Because of the prompting, then, semi-structured interviewing techniques can ensure that a central set of issues are covered by each respondent – this gives the chance for a more systematic analysis than might be possible with an unstructured interview. At the same time, respondents still have the opportunity to raise issues that are of particular importance to them.

Structured interviews ask the respondents to choose a response from within a pre-set range. This might mean, for example, asking people to rate certain aspects of the product on a numerical scale, or asking them to choose a response or responses from a set of categories. Again, considering the case of a high-end television, this could mean marking against items on a list to indicate which aspects of the product's design they particularly liked or disliked. The responses from these interviews lend themselves to structured quantitative analysis. However, in order to be able to predetermine the possible response categories, the investigator must have a fairly clear idea of the issues that need investigation.

Advantages

Interviews are a versatile method in that they can be used throughout the design process. As with questionnaires, questions can be formulated for investigating people characteristics and people's attitudes towards proto-types as well as finished products.

Because the investigator administers the interview directly to the respondents, the likelihood of the respondents misinterpreting the question to which they are replying is less than those associated with a questionnaire. With a questionnaire, the respondent has to make an interpretation of the question based purely on what is written down – if this is misinterpreted, he or she may not be able to give a meaningful answer to the question as intended by the investigator. In an interview situation, however, the respondent is free to ask the investigator about anything that he or she is unsure about. Similarly, if the respondent replies in a way that is not meaningful in the context of the question as intended by the investigator, then the investigator can rephrase the question in a way that the respondent will understand. The interactive nature of an interview, then, can potentially make the data gathered more valid than that from questionnaires.

Another way of considering this advantage is to trade off the validity of the data gathered against the time that goes into preparing the investigation instrument. Given that a certain level of accuracy may be required, it should be possible to achieve this with less preparation effort with an interview than would be necessary with a questionnaire. When preparing a questionnaire, the investigator must be sure that questions are worded without ambiguity and that the nature of the replies required by respondents is clear. With an interview, conversely, it may be possible to compensate for some deficiencies in question formulation by the two-way communication during the interview session itself.

Another advantage that an interview can have over a questionnaire is the lesser extent to which the respondents are self-selecting. With question-naires, there is often a low return rate. This is a problem because those most likely to return questionnaires may be those with unrepresentative opinions about a certain issue. In the context of evaluating the pleasurability of a product, this might be those with particularly strongly negative or, perhaps, strongly positive views about a product's pleasurability. An analysis based on the sample returned might, then, give a distorted view of how people in general would relate to a product.

With interviews, there is still a degree to which respondents are self-selecting; after all, those taking part in interviews must be asked for their consent in advance and there is no guarantee that everyone approached will be willing to give up their time for this. However, once someone has agreed to take part, it would be unusual for him or her not to go on and complete the interview session. This is unlike what may happen with questionnaires, where even those originally intending to fill them in may end up leaving them half-finished or may never get round to starting them at all.

Disadvantages

The costs of administering a series of interviews are high in comparison with collating information from a similar number of questionnaires. This is because, with an interview, the investigator will need to be present in order to ask the questions, whilst, with a questionnaire, the respondents would normally answer without the investigator present. Where a large number of respondents are required, this can be very expensive in terms of investigator time.

Another disadvantage of having the investigator present is the risk of having the data gathered distorted by an investigator/respondent effect. Whilst data gathered from questionnaires may give an unrepresentatively extreme view of user opinion, the opinions given in interviews could be unrepresentatively moderate. When giving opinions to another person, it is possible that respondents may not want to give vent to particularly strongly held views, which they might have felt more comfortable expressing when given the anonymity afforded by a questionnaire. This is because, when interacting with others, there may be a desire to be seen as being 'pleasant' and 'reasonable'. Perhaps respondents will fear that the interviewer might find them unreasonable or unpleasant if the answers they give to questions are too extreme – particularly, of course, if they are very negative.

Immersion

This technique involves the investigator experiencing a product himself or herself and evaluating the pleasurability of the product on the basis of their own experiences. The technique might be seen as semi-empirical. On the one hand, it involves no participants other than the investigator. On the other hand, the investigator is making judgements about the product's pleasurability or displeasurability based on his or her experience of it, rather than solely on the basis of his or her expertise or by evaluation against a checklist of product properties.

Usually, immersion will involve the investigator experiencing a finished product over a period of time. If, for example, the product under investigation were a vacuum cleaner, then the investigator might use this in his or her home. Each time he or she used it, the investigator would record his or her experiences and the opinions that he or she has about the product. He or she might also record his or her impressions when he or she first saw the product and the reactions of others to the product.

If, for example, the product under investigation were a mobile telephone, the investigator might begin by going to an electrical or electronics retailer and looking at the product on the shelf next to other mobile phones. He or she might make observations about how this phone compares with others on display and what his or her reaction to the product was when first seeing it. Would it be the product that he or she would have chosen to buy? If so, why? If not, why not? In what way does his or her reaction to this product differ from his or her reactions to the other phones? Does the design of the phone under evaluation seem more or less sophisticated than the others on display? Does it look like a high-status product or does it look 'cheap and cheerful' or even 'cheap and nasty'? These are examples of the sorts of judgements the investigator may make.

The investigator might then record his or her impressions on first picking up the phone. Examples of the issues to be addressed might include the feel of the phone in the hand – is it pleasant to hold, and does the feeling of the phone radiate an impression of high quality? Do the buttons feel pleasant to the touch? When they are pressed do they give clear and reassuring tactile feedback? Does the phone fit easily into the pocket of the investigator's jacket? Is it comfortable to walk around with it in the pocket?

Next, the investigator might look at some of the usability aspects of the phone. He or she might begin by trying few basic tasks – making a phone call, receiving a phone call, putting a number into the memory or selecting a number from those stored in the memory. After this, he or she might go 'out and about' with the phone – using it on the street, in the train, in the home, in restaurants and bars or in the workplace. How does he or she feel when using the phone in these situations – embarrassed, proud, important or idiotic? What role does the design of the phone have in affecting the way the investigator feels?

Perhaps the investigator might try using the phone under more 'extreme' circumstances – during a hike in the countryside, when driving (not a wise thing to do, but people unfortunately do use mobile phones in this context, so it may be sensible to check the effects of this) or at a football game. Are there any particular aspects of the design of the phone that affect how pleasurable it is in such circumstances? What happens if the phone is dropped? If it is chipped or scratched does this spoil the appearance of the phone, or does it 'wear well' – retaining its appeal despite sustaining wear or damage?

After a period of experience with the phone, the investigator will then make a judgement about how well the phone performs against the criteria by

162

which its pleasurability is to be judged. In the case of a particular phone, for example, these could be as listed below.

- The phone should be guessable. It should be usable at the first or second time of trying for the most basic tasks.
- The phone should be sensorially pleasing in terms of its tactile aspects. It should be pleasant to hold in the hand and the buttons on the keypad should feel pleasant to the touch.
- The phone should be a status symbol. The person using the phone should feel that being seen with it enhances his or her status, and others should respond positively when seeing it or when remarking upon it.
- The phone should be easily portable. It should fit nicely into a jacket pocket, handbag or briefcase.

Advantages

As with the non-empirical methods, not requiring participants other than the investigator himself or herself can be convenient and it preserves confidentiality.

Perhaps the biggest advantage of this technique is that it gives the investigator a first-hand insight as to what it is really like to experience a product. This makes immersion unique amongst the techniques outlined in this chapter. With the other empirical methods the investigator is observing or being told about the experience of others, whilst with non-empirical techniques the investigator is simply making a prediction or listening to the predictions of others about how pleasurable or displeasurable a product may be.

First-hand experience can be extremely valuable as it cuts out the potential miscommunications or misunderstandings inherent in observing or receiving feedback from others. Perhaps more importantly, the investigator can gain a sense of empathy with others who experience the product. This can make the technique extremely valuable when used in combination with other methods. For example, if on, say, a questionnaire, a respondent has indicated that a product is 'cumbersome to carry around' then the investigator can get a real feeling for what this means. He or she can also gain insights into exactly what it is about the product's design that makes it cumbersome.

Most empirical techniques, by their very nature, tend to be more diagnostic than directly prescriptive. Whist participants may be able to report their experiences accurately, they may have far greater difficulty in making practical suggestions about how the product can be improved. After all, investigation participants will not usually have any specialist design knowledge, so it might be unrealistic to expect them to give workable design solutions. With immersion, on the other hand, the investigator will usually be a human-factors specialist: someone who is charged with giving, and is qualified to give, guidance in the design process in order to improve the product. It should, then, be possible for him or her to derive design solutions directly from his or her experiences.

Disadvantages

Perhaps the most obvious potential weakness of this method is that the evaluation depends on the experiences of only one person – the investigator – who may not be, and indeed is unlikely to be, representative of the people for whom the product is designed. The effectiveness of the method, then, is likely to be highly dependent on the ability of the investigator to empathise with those in the product's target group.

Returning to the example of the mobile phone, it may be, for example, that the phone was aimed at 'yuppies', a description that may not fit the investigator. The investigator might, for example, have no interest in products as status symbols. However, in order to conduct the investigation effectively, he or she must be able to empathise with yuppies and understand how someone who regarded himself or herself as a yuppie might want to be perceived by others. Some investigators may be more empathetic than others, so there may be an effect for the abilities of the investigator in this respect.

Another, perhaps even more significant way in which the investigator might differ from those for whom the product is designed is that the investigator may have a specialist interest in the product under investigation. This might encourage him or her to relate to the product in a manner that differs significantly from the way in which the target group would relate to it. Indeed, this is likely. After all, the investigator is likely to be a human-factors specialist from the company whose product is under evaluation. This might mean that the sorts of things that would make a product particularly pleasurable or displeasurable for the investigator might be very different from those for whom the product was designed.

For example, most human-factors specialists will be steeped in a tradition of looking at usability and interface design issues. It may be, then, that when experiencing a product, they are particularly likely to have a strong reaction to usability issues. These may or may not be of great interest to the product's target group. It is important, then, that the investigator does not let his or her personal agenda colour the overall assessment of the product. Similarly, an investigator's knowledge of a particular product or product type might give them special insights into the product that will affect his or her experience of it. For example, his or her knowledge of how a product works might mean that functions that gave problems in use to others may be easy for the investigator to use. Again, then, this is a way in which the effectiveness of the method is likely to be influenced by the ability of the investigator to 'look outside' himself or herself.

Laddering

Laddering – a method with a history of use in marketing – can be used to understand the links between formal product properties, experiential product properties, product benefits and the characteristics of the person experiencing the product. This technique can be very useful in the context of pleasure-based approaches to human factors, as it can provide information that will be valuable throughout the whole of the product-creation process.

The basis of the technique is that a investigator asks a participant to mention some feature of a product that he or she feels particularly positive or negative about. The investigator then asks why. The participant makes a statement in response and the investigator then asks why again. This question/response cycle continues until the participant is unable to give a reasoned response. Once this has happened, it might be assumed that the participant has revealed one of his or her fundamental values. The investigator will then ask about another aspect of the product and the process will be repeated until eventually the participant cannot think of another positive or negative feature of the product.

To illustrate how the technique works, consider the following example. Imagine that the subject under discussion were a soft drink – in this case a diet cola.

Investigator:	Please tell me something that you like about this product.
Participant:	I like that there is only one calorie per glass.
Investigator:	Why?
Participant:	Because that means the drink is not fattening.
Investigator:	Why is that important?
Participant:	Because it means that I won't get fat if I drink this product.
Investigator:	Why is that important?
Participant:	Because it is not nice to be fat.
Investigator:	Why is it not nice to be fat?
Participant:	Because it's more beautiful to be thin.
Investigator:	Why is it important to be beautiful?
Participant:	Because people respond to you more positively.
Investigator:	Why is it important for people to respond to you positively?
Participant:	It makes life easier.
Investigator:	Why is it important to have an easy life?
Participant:	It just is.

In this case, then, the ladder would be as given in Figure 4.3.

In terms of the properties and benefits of the product and the characteristics of the person, this reveals the following information: the formal property 'one calorie' is associated with the experiential property 'non-fattening'. The participant regards being thin as being associated with being

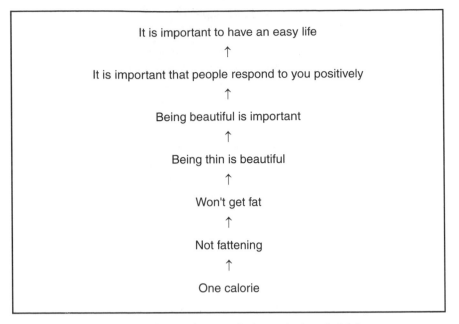

Figure 4.3 Possible ladder from a feature of a low-calorie soft drink

beautiful – so the contribution of the drink to helping the person to be beautiful is the associated benefit. This benefit is important to this participant as it helps to contribute to having an easy life – a high-level goal of this person.

Now consider another example. This time imagine that a participant is being asked to consider a consumer electronics product – in this case a stereo system. Imagine that they are asked to identify something about the product that they particularly dislike and in this case they dislike the colour. The dialogue might be as follows.

Investigator:	Please tell me something that you don't like about this product.
Participant:	I don't like the colour.
Investigator:	Why don't you like the colour?
Participant:	I don't like black on stereos.
Investigator:	Why don't you like black-coloured stereos?
Participant:	All stereos are black.
Investigator:	Why is it a bad thing that this is the same colour as other stereos?
Participant:	Because it's boring. I want my stereo to be different.
Investigator:	Why do you want to have a stereo that is different from others?

Participant:	I want to be able to choose something that expresses my own tastes.
Investigator:	Why do you want to be able to express your own tastes?
Participant:	I want to be an individual, not just go along with the crowd.
Investigator:	Why do you want to be an individual?
Participant:	I just do.

The ladder resulting from this dialogue is summarised in Figure 4.4.

In this case, then, the following has been learned about the product and the person. The formal property 'black' is associated, in the case of this participant, with the experiential property 'boring'. The participant regards expressing his or her individuality as important and sees his or her choice of stereo as being a factor in expressing individuality. The associated drawback, then, is that the colour of the stereo is not helping the person to express his or her individuality.

Advantages

A major advantage of laddering is that it is a method that enables the investigator to gather information about formal properties, experiential properties, desired benefits and the characteristics of the people for whom the product is designed. It also gives information about the relation between these aspects. This means that properties can be linked directly to benefits and that these benefits can, in turn, be linked directly to the characteristics of the target group.

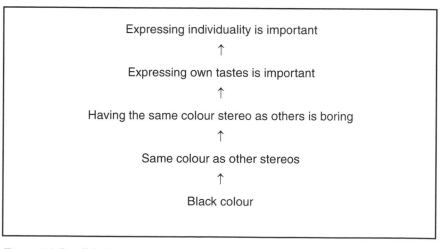

Figure 4.4 Possible ladder from a feature of a stereo system

167

The method can be used at any point in the product-creation process. The participant might be asked to comment on prototypes or finished products.

Disadvantages

A disadvantage associated with laddering is that the method can be comparatively time consuming, both for the investigator and for the participant. Unlike, for example, questionnaires and experience diaries, the method cannot be implemented remote from the presence of the investigator and only one person can participate at a time.

Another potential disadvantage is that the method can be somewhat demanding of participants. They are asked a continual stream of 'why?' questions. A potential danger of this is that they may feel pressured to keep on justifying their statements with rational explanations. This may be a problem because it may mislead the investigator as to which are really the highest-level values of the participants. Consider again the examples given above. In the case of the low-calorie soft drink, the high-level value of the participant that was established from the session was that he or she felt that living an easy life was important. On the basis of this, the manufacturers might be led to believe that promoting the drink with images of, for example, people relaxing might be the most effective way of making the drink appealing to people such as this participant.

It could be, however, that, in reality, the participant was only concerned with looking beautiful and that his or her responses to the subsequent questions were simply 'rationalisations' of this value rather than reflections of his or her true feelings. Were this so, then the images of people relaxing may not be the most appropriate for promotion of the product. Rather, the target group might look more favourably on the drink were it to be promoted via images of beautiful people drinking it.

Similarly, in the case of the discussion about the stereo system, imagine that the participant simply didn't like black, for no other reason than that it offended his or her aesthetic sensibilities. The subsequent replies, leading to the conclusion that the participant valued the expression of his or her individuality, may have been misleading rationalisations. This may lead to the creation of inappropriate design directions. For example, on the basis of the laddering session, the investigator might recommend that the design should have an extreme form, making it look very different from other stereos on the market. In reality, however, such a form might be off-putting for those for whom the product is designed. They may simply prefer a conventional design in some colour other than black.

Participative creation

Participative creation involves a group of participants gathered together, with, for example, designers and human-factors specialists, in order to discuss issues relating to a product's design. This might mean simply listing their requirements in terms of the benefits that they would hope to gain from a product – helping to set the product benefits specification. It may, however, include discussion of the aesthetic and functional properties that a product should have and even helping to sketch out some ideas for possible designs.

Participative creation sessions differ from focus groups primarily in that they involve users in a 'hands-on' way, rather than simply asking them to discuss issues. For example, in a study investigating design for different cultures (Hartevelt and van Vianen 1994) two participative creation sessions were conducted – one with Japanese people and the other with Europeans. The participants in each session discussed their requirements for television sets. This included discussion of the context in which they used their televisions and the design issues that had an influence on their purchase decisions with respect to televisions. They were also shown several television sets and asked to comment on the design properties of each. Following this, they were involved in a co-design session in which they gave suggestions for the design of a new television.

Advantages

Participative creation represents a very direct way of getting people involved right from the start of the design process. Not only are participants asked what their requirements are, but they can also become involved in translating these into design solutions. Having people work with designers to sketch out parts of a product can be helpful, as it exposes designers directly to those that they are designing for, rather than having these people's requirements communicated to them indirectly, for example via a human-factors specialist.

Disadvantages

Participative creation can be comparatively demanding for participants – both in terms of the time that the session takes – for example, participants in Hartevelt and van Vianen's sessions each gave over three hours of their time – and the amount of work that being a participant can entail. This may make it difficult to find participants, unless there are people who are particularly motivated to attend the sessions and/or who have a large amount of time available. It is also questionable as to whether it is advisable to get participants so directly involved in the creation of design solutions. After all, they are not designers and it might be unrealistic to expect participants to come up with sensible and feasible design solutions.

Although direct communication between people and designers was cited as an advantage of participative creation, there may also be disadvantages to this. It may be, for example, that if designers are present, the people participating feel restricted with respect to communicating their ideas. It could be, for example, that the participants might feel embarrassed about saying things that they felt that designers would regard as 'foolish'. This, then, might lead to incomplete or inaccurate information as to the needs and wishes of those for whom the product is being designed.

Controlled observation

A controlled observation is a formally designed investigation with comparatively tight controls and balances. In contrast to field observations, the aim is not to understand the real-life context in which a product is experienced, but rather to isolate the effects of particular design decisions on the overall pleasurability of the product – this requires the gathering of comparatively noise-free data.

For example, in an evaluation of the user friendliness of a product, the investigator might set the participants a number of tasks to do with a product or a working prototype. In such a case, there will typically be balances on the order in which tasks are set, in order to minimise the possible effects of knowledge transfer between tasks. Consider, for example, using a word-processing package for formatting text and for altering fonts. It might be that formatting tasks are done by highlighting the text to be formatted and then selecting the appropriate command from a menu. To alter the fonts, it might be that the text whose font is to be altered must be selected and then a command for the font required selected from a menu. There would, then, be similarity here between these two types of task. Both require highlighting text and then selecting commands from a menu.

If participants were always to be set formatting tasks first and were asked to alter fonts later in the investigation session, then they might be able to guess how to alter the fonts on the basis of their experience of the formatting tasks. Even if they had struggled with the formatting tasks, then, the experience gained from completing these tasks may mean that when they subsequently came to alter the fonts this presented no problem. When looking at the data from an investigation designed without balances for task ordering, this may give the impression that formatting was a difficult task that caused a lot of problems, but that altering fonts was an easy task that people had little difficulty in completing at the first attempt. This might lead to the conclusion that the properties of the interaction design were appropriate with respect to altering fonts, yet inappropriate with respect to formatting tasks. Of course, this conclusion is almost certainly erroneous. After all, the two tasks are performed in similar ways.

Task-ordering balances should eliminate any such effects. For example, half the participants could be set the formatting tasks before the font tasks and the other half the font tasks before the formatting tasks. This should mean, then, that for half the participants learning effects from the formatting tasks should be carried to the font tasks and for the other half effects should be carried the other way – thus balancing out. Overall, it is likely that mean performance should be similar for each task.

As their name suggests, controlled observations are carried out under tightly controlled conditions. This means, for example, eliminating distractions that could potentially interfere with the way in which the participant responds to the product. Potential distractions could come from the environment in which the investigation is conducted – for example, the sound of others having conversations, or distractions in the participants' field of vision. They could also come from having to do other tasks at the same time as using the product under investigation. In order to eliminate these effects, controlled observations are often conducted in somewhat 'sterile' laboratory environments in accordance with fairly rigid protocols.

Advantages

Because the data gathered from controlled observations is comparatively 'pure', the method is good for picking up relatively small effects that might not be detectable with other methods, where there is the possibility that they will be swamped by noise or confounded by other effects. This can make the controlled observation an effective method for investigating specific design options by direct comparison. For example, in the case of a software package, two prototypes could be constructed that differ with respect to some specific aspect – for example, the colour combination used in the on-screen display. By applying the correct controls and balances, it should be possible to gather data that validly indicates which of the colour combinations draws the more positive response from the participants. In this case, for example, facial expressions might be monitored in order to make judgements about which combination did so.

Another advantage gained from gathering comparatively pure quantitative data is that it can be used as material for inferential tests for statistical significance. This, then, can give a comparatively unambiguous indication of whether there are any apparent effects in user response to one design over another.

Disadvantages

Perhaps the main disadvantage of controlled observations is that in order to achieve the levels of control and balance necessary to keep the data free of noise and other effects, the environment and circumstances in which these observations are conducted often tend to be somewhat artificial. There may be no guarantee, then, that effects which appear to be highly significant, judging by the data collected in controlled observations, will actually prove to be important when the product is experienced in a real-life context.

Another possible problem associated with conducting investigations in very artificial environments is that the artificiality may affect the way in which the participants experience the products or concepts. It is possible, for example, that aspects of the design that appear favourable in a laboratory situation may not be so favourable in the real-life context in which the product is likely to be experienced.

Consider, for example, an investigation of the aesthetic properties of two different design concepts for a home entertainment product, such as a stereo. If a participant preferred the aesthetics of concept A over concept B when comparing them under laboratory conditions, this would not necessarily mean that he or she would prefer concept A if the products were to be placed in his or her living room. Perhaps, for example, concept A might be the more attention grabbing of the two concepts. This property might make concept A more appealing when viewed in an isolated context, yet it might, for example, lead to aesthetic clashes when the product were placed in the interior of the person's home.

Non-empirical methods

Expert appraisal

Here a product is evaluated on the basis of whether an 'expert' or 'experts' regard it as being designed in such a way that it will be pleasurable. An expert, in this context, is an investigator whose education, professional training and experience make him or her able to make an informed judgement on issues with respect to the product or concept under investigation. Perhaps the expert will have specialist knowledge about the people for whom a product is designed. In the context of, for example, a product designed for blind people, the investigator might be someone with a specialist knowledge of design for the blind or of issues associated with the day-to-day difficulties that can be associated with blindness.

Alternatively, the expert may be someone with specialist knowledge of a particular aspect of design or of a particular application area. For example, in the context of evaluating, say, a computer-based software package, then the expert might be someone with a background in human–computer interaction (HCI) and a specialist knowledge of the type of program under investigation.

The sorts of issues covered may be similar to those covered by someone using a property checklist, although with expert appraisal the range of issues looked at may be narrower with the investigator going into more depth. This is because his or her expert knowledge should give the investigator an idea of which the really important issues are in a particular context, as well as an idea of the details that can really make a difference to the pleasurability of a product of a particular type.

Sometimes more than one expert may give an opinion on a product. They may work together to give their assessment, or they may rate the product separately. The latter approach supports a degree of cross-validation, through comparison of the assessments given by the different investigators.

Advantages

As with all non-empirical methods, no participants are needed and expert appraisal is also a good method for providing diagnostic and prescriptive analyses. The identification of an issue with a likely effect on product plea-surability will be based on a particular characteristic or particular characteristics of a design. The investigator's specialist knowledge should lead directly to solutions as to how the design can be improved.

Suppose, for example, that the product under investigation were a soft-ware-based information retrieval system. If the system required that the users type in long command strings, the investigator might predict that this will possibly create usability problems due to the users forgetting or mis-typing strings. This could lead directly to a number of potential solutions. For example, the investigator could recommend that shorter and more memorable command strings were used or that the system was redesigned as a menu-based interface. Similarly, if the investigator felt that the colour used in the interface carried the wrong emotional message, then he or she could recommend a colour that would be more appropriate.

Disadvantages

As with all non-empirical techniques there is no direct evidence from users that any of the issues identified by the investigator as affecting product plea-surability are actually of any importance. For example, in the context of expert evaluations of product interaction style, users can confound expert expectations by adapting to what might seem to be major shortcomings in a product or by being tripped up by a fault that appeared trivial to the investi-gator. Similarly, an aesthetic property that the investigator expected to trigger a particularly strong reaction may not be that important to the people for whom the product is designed, whilst a seemingly minor aesthetic detail might trigger a very strong reaction. In the absence of empirical data, then, the method is totally reliant on the expertise of an individual investi-gator or of a group of investigators.

For example, Kerr and Jordan (1994) report an evaluation of the inter-face to a software-based prototype telephone system in which two HCI experts were consulted in order to make predictions about the suitability of functional groupings on the telephone and their likely effects on usability. They found that, of twelve predictions made by the experts, only five were supported by task performance data in a subsequent empirical evaluation. By contrast, eleven out of twelve predictions made on the basis of data gathered from potential users – using a questionnaire-based tool designed by Kerr and Jordan – were supported. This effect may also apply to aspects of pleasurability outside of product usability.

Property checklists

In the context of usability evaluation, property checklists list a series of design properties that, according to accepted human-factors 'wisdom', will affect whether a product is usable. Usually, these will state the high-level properties of usable design, such as consistency, compatibility, good feedback, etc. They will then list low-level design issues relating to these – these might be on the level of, say, the height of characters on a computer screen or the height of characters on labels on products, or specifying the position of displays and controls. The idea is that the investigator checks the product being evaluated to see whether its design conforms to the properties on the list. Where it does not usability problems might be expected.

A good example of a property checklist is contained in Ravden and Johnson's book, *Evaluating Usability of Human–Computer Interfaces: A Practical Method* (1989). Originally designed for the evaluation of human–computer interfaces, this checklist has also been used as a basis for the evaluation of other types of product (Johnson 1996).

Unfortunately, in the context of measuring other aspects of product pleasurability such 'off-the-shelf' diagnostic checklists do not exist. This is probably a function of the newness of these wider pleasure issues within the human-factors discipline. There is not yet, for example, a sufficiently large body of experience and literature on the issue to be confident about generalisable links between specific properties of a product and its pleasurability. Consider, for example, the labelling on a display or button. From a usability point of view it may be possible to specify the height and contrast of the characters that are required in order that the label be legible from a particular distance, under particular lighting conditions or when a particular font style is used.

It would not – certainly not yet – be possible to make such detailed, low-level, generalisable assertions about what would make a product pleasurable in the wider sense. For now, at least, generalisable property checklists for evaluating pleasure tend to consist of lists of design elements for the investigator to assess in terms of their anticipated contribution to product pleasurability. Imagine, for example, that the product under evaluation were a power-drill. Imagine that the product had been designed to bring the following benefits: control, power, robustness and safety. The investigator would then look at the various elements of the product design – form, colour, graphics, etc. – to see if, between them, they emphasised these values. A checklist for evaluation of such a product might be as given in Figure 4.5.

Product element	Evaluation criteria
Form	Does the form of the product emphasise control? Does the form of the product emphasise power? Does the form of the product emphasise robustness? Does the form of the product emphasise safety?
Colour	Does the colour of the product emphasise control? Does the colour of the product emphasise power? Does the colour of the product emphasise robustness? Does the colour of the product emphasise safety?
Materials and finishings	Do the materials and finishings of the product emphasise control? Do the materials and finishings of the product emphasise power? Do the materials and finishings of the product emphasise robustness? Do the materials and finishings of the product emphasise safety?
Graphics	Do the graphics on the product emphasise control? Do the graphics on the product emphasise power? Do the graphics on the product emphasise robustness? Do the graphics on the product emphasise safety?
Sound	Does the sound of the product emphasise control? Does the sound of the product emphasise power? Does the sound of the product emphasise robustness? Does the sound of the product emphasise safety?
Functionality	Does the functionality of the product emphasise control? Does the functionality of the product emphasise power? Does the functionality of the product emphasise robustness? Does the functionality of the product emphasise safety?
Interaction design	Does the interaction design of the product emphasise control? Does the interaction design of the product emphasise power? Does the interaction design of the product emphasise robustness? Does the interaction design of the product emphasise safety?

Figure 4.5   Property checklist for evaluation of a power-drill

In essence, then, property checklists for measuring product pleasurability may be little more than a means of structuring an expert appraisal. After all, the checklist shown in Figure 4.5 merely takes the investigator through the design element by element and gives no guidance as to the properties that each element should have in order to fulfil the evaluation criteria – that is left to the investigator's expert judgement.

However, this is not always the case. If, for example, one of the aspects of pleasurableness on which the product were to be judged were its usability, then a detailed checklist such as Ravden and Johnson's (1989) could be employed. Similarly, if a criterion were that the product should be helpful in performing a particular range of tasks, then the checklist might include detailed items relating to functionality.

Advantages

As with all non-empirical methods, not having participants can be convenient and it preserves confidentiality.

Another advantage of this technique is that, where more detailed 'low-level' items are included in the checklist – for example, in the context of evaluating interaction design – it can lead directly to design solutions. Indeed, the criteria against which the product is being judged can be indicative of what the design solutions should be. As a simple example, consider the legibility of textual button labels. If, for example, the labels had to be legible at a distance of 1.5 metres, then the criterion on the checklist might state that characters should be at least, say, 6 millimetres high. This, then, is not only giving a criterion against which to evaluate, but also a design solution if the criterion is not met.

Property checklists can be used throughout the design process. Right from the start, the criteria that they list can be used as part of requirements capture and product property specification. They can also be used for evaluation of visual and functional prototypes, as well as the evaluation of finished products.

Disadvantages

The validity of an investigation conducted using a property checklist is dependent on the accuracy of 'expert' judgement. This is so in two senses. First, it is dependent on the judgement of the person or people who compiled the checklist in the first place. Whilst some of the items on the list may represent design criteria that have been established on the basis of years of human-factors research, others may be more speculative, particularly if those compiling the checklist have tried to set low-level criteria against which to judge some of the more ephemeral aspects of product pleasurability. So, for example, there has indeed been much research into the size of characters required for legibility at certain distances. There is now, then, more or less a consensus as to what is required here. Consider, conversely, the properties associated with a benefit such as emphasising a person's masculinity or femininity. Within the literature, there is still very little on this issue – for one of the few texts on the subject see *The Gendered Object* (Kirkham 1996) – so any assertions made about the formal properties associated with this will probably owe a great deal to the checklist compiler's personal judgement. It may be, then, that criteria listed on a property checklist for judging the extent to which a product communicates masculinity or femininity would be less reliable.

The second expert upon whom the validity of a property checklist-based evaluation may be dependent is the investigator. Even where the formal properties connected with a particular experiential property or product benefit are well-established, there may be complications. For example, when analysing the legibility of the graphics on a product, subtle variations in font style may or may not have an effect. If the fonts used on the product differ from those given in the checklist then the investigator will have to make a personal judgement about whether or not this will affect legibility. Where the checklist has little or nothing to say about the low-level criteria against which the product should be judged, then the evaluation of the product's design will be almost entirely dependent on the judgement of the investigator.

Another disadvantage of property checklists is that it is not always possible to judge how big an effect any deviation from the listed properties will have. When people actually come to experience products, they may be indifferent to or unaffected by certain aspects of the product's design, yet highly affected by others. The danger is, then, that, without actually observing people experiencing products, it may be difficult to tell which aspects will prove critical to overall product pleasurability. This can cause problems when making design recommendations. Because of the limited time and budgets that are usually available to those undertaking such work, it will normally be necessary to have a clear idea of what the priorities for attention are. This is difficult to do without a clear idea of the comparative contribution of the design elements to overall product pleasurability. This difficulty in estimating the comparative importance of the properties of product elements is, perhaps, a problem that could be said to be associated with all non-empirical evaluation methods.

Examples of application

So far in this chapter a number of methods for use in the creation and evaluation of pleasurable products have been outlined. In the remainder of this chapter examples of the application of some of these methods will be given. These examples are drawn from the small but expanding literature reporting pleasure-based approaches to human factors.

Kansei Engineering

Kansei Engineering was developed at the University of Hiroshima, mainly through the work of Mitsuo Nagamachi. Nagamachi details the fundamentals of this technique and the history of its development in *The Story of Kansei Engineering* (1995). Kansei Engineering – which translates roughly as 'Pleasure Engineering' – helps the investigator to understand the relationship between the formal properties and experiential properties of a product. It can also be used as a means of gaining an insight into the sorts of benefits that people wish to gain from products and the product properties via which these benefits can be delivered.

Kansei Engineering can work in two ways – Nagamachi (1995) refers to these as the two directions of 'flow'. One direction of flow is termed 'from design to diagnosis'. This involves manipulating individual aspects of a product's formal properties in order to test the effect of the alteration on users' overall response to the product. This technique has been used to assist in the design of a diverse range of products. Nagamachi (1997) describes examples that range from automobiles through camcorders to clothing.

This technique is probably best described by means of an example. A case study – reported by Ishihara *et al.* (1997) – is summarised below, demonstrating how Kansei Engineering was applied to the design of cans for coffee powder. The method used in this direction of flow is structured interviewing that leads to the gathering of quantitative data describing people's responses to the product's formal properties. This data is then treated with statistical analyses in order to establish the relationship between these formal properties and the experiential properties of the product.

The other direction of flow is from context to design. This involves looking at the scenarios and contexts in which the product is used and then drawing conclusions about the implications of this for the design. This second direction of flow involves the gathering of qualitative data via field observations. In this case, the data is used to help establish the link between the formal properties of a design and the benefits associated with the product. Again, a case study is reported below, demonstrating the application of Kansei Engineering in this direction of flow.

Kansei flow 1: from design to diagnosis

The application of Kansei Engineering in this direction of flow has been used to assist in the design of a diverse range of products. Nagamachi (1997) describes examples that range from automobiles to brassières. The following case study reports how Kansei Engineering was applied to the design of cans for coffee powder.

In this study, seventy-two alternative designs of coffee can, differing in terms of combinations of formal properties, were presented to a panel of ten subjects. Examples of the types of formal properties that differed from concept to concept were the size of the logos on the cans, the style of the fonts used in labelling, the types of graphic imagery used and the colours of the cans. Each member of the panel was asked to rate each of the designs according to how they fitted with a series of descriptor adjectives. There were eighty-six of these descriptor adjectives. These descriptor adjectives represented potential experiential properties of the designs. Examples were: showy, calm, masculine, feminine, soft, individual, high grade, sweet, milky, etc. Panellists rated each of the seventy-two designs according to the eighty-six descriptor adjectives by marking five-point numerical scales to indicate the degree to which they felt each of the designs exhibited each of these experiential properties.

A cluster analysis was then carried out in order to establish the links between the formal and experiential properties of the designs. For example, one cluster of cans emerged that were regarded by the panellists as having the experiential properties 'milky', 'soft' and 'sweet' – this cluster of cans was characterised in terms of formal properties by the use of beige colouring for the majority of their surface. Another cluster was seen as having the experiential properties 'masculine', 'adult' and 'strong' – these experiential properties were all associated with the formal property of having a large logo on the cans. A third example of a cluster was one that was seen experientially as being 'unique', 'sporty' and 'individual' – this was related to the formal property of blue and white colouring in the designs.

Kansei flow 2: from context to design

The other 'direction' in which Kansei Engineering can be applied is to observe products being experienced in the 'natural' environment and context of their use. The qualitative data gathered from these field observations is then used in order to give insights into the sorts of benefits that people may want from a particular product and the formal properties associated with such benefits. These approaches can often mirror the traditional human-factors field study. Indeed, sometimes, the studies will involve little more than looking at the ergonomics of product use in a particular context. Often, however, the emotional aspects of product use will also be considered. Nagamachi (1997) reports on this type of approach in the context of the design of three different products – a refrigerator, a camcorder and a brassière.

The refrigerator project, carried out for Sharp, represented the first successful application of Kansei Engineering to product design (Nagamachi 1995). In essence this was little or nothing more than a field-based approach to user-centred design with traditional ergonomic concerns at the centre. Nagamachi and his associates visited the homes of a number of a refrigerator users in order to investigate the issues that arose in use. These mainly involved physical issues such as users having to bend awkwardly in order to reach particular items in the fridge. The recommendation, basically, was to alter the positions of the various components of the fridge: nothing different, then, from traditional usability-based approaches.

The approach to camcorder design was based on a rich understanding of the context in which people tended to use the product. From looking at marketing data Nagamachi discovered that the biggest single user group for camcorders was families who had just had babies. This meant, for example, that users had to get down on their hands and knees in order to film a baby crawling. To solve this problem Nagamachi recommended a lens that could be swivelled through 360 degrees. Another aspect of use that Nagamachi noticed was that users would often enjoy showing others what they had just filmed. In order to show this, the users would ask their friends to look through the viewfinder and would then play back what they had just filmed. To improve on this rather inelegant solution Nagamachi proposed adding a large, bright monitor to the back of the camcorder. This enabled others to view the playback without having to look through the viewfinder. In this case, then, Nagamachi established that a benefit that people might want from the camera was the ability to show their friends what they had just filmed. His suggestion was that this benefit could be delivered via the formal properties of the monitor.

The approach to the brassière design, on the other hand, relied on interviews – in this case unstructured interviews. The approach here, to some extent, mirrored that used in the study of the coffee can design. However, whilst in the coffee can study participants were presented with a number of concepts and asked to rate them with respect to pre-identified descriptors reflecting potential experiential properties, the participants in the brassière study were asked to identify the benefits that should be delivered via the design of the brassière. Five-hundred women were asked about the issues that were important in terms of how wearing the brassière should make them feel. 'Beautiful' and 'graceful' emerged as the two most important benefits and subsequent design concepts were evaluated according to how they made their wearers feel with respect to these benefits.

Nagamachi has developed databases to assist designers in using Kansei Engineering. These databases contain information about links that have been established between formal properties of products and people's reactions to those products. In addition, images of designs that are associated with particular user reactions are stored in the database. When a designer wishes to create a new concept, he or she can feed information into the database about the sort of benefits he or she is hoping to deliver to the people for whom the product is being designed. The database then returns a design suggestion in the form of an image of a product containing the formal properties required in order to deliver the desired benefit. Nagamachi (1995) reports on the application of computer-supported Kansei Engineering to the design of a car steering wheel.

First, Nagamachi and his team analysed dialogues between car salespersons and customers in car show rooms in order to gather information about the different sorts of ways in which people might react to new cars and the sorts of benefits that may be looked for. This was supported by an analysis of motoring magazines – looking at the words that journalists used to describe their reactions to new cars. Nagamachi's team used the outcomes of these two studies in order to derive a list of benefits that could be associated with new cars and the experiential properties associated with these benefits. These experiential properties were then used as the basis for creating semantic differential scales that would be used in the Kansei analysis.

Fifty-nine samples of steering-wheel design – differing in terms of a number of formal properties – were then collected from a variety of cars and were photographed to make slides. These slides were then shown to fifty men and women, who evaluated each according to the potential experiential properties of the design, which were listed on the semantic differential scales.

Each of the steering-wheel designs was classified according to thirteen different formal properties. These included, for example, the number of spokes, the size of the pad in the centre of the wheel and the thickness of the spokes. A cluster analysis was then carried out in order to establish the links between the formal and experiential product properties. These links were then fed into databases. One database pertained to the formal-experiential property links made by the women and another to the formal-experiential property links made by the men, while a third database contained the combined data from the men and women. The designer was then able to consult the database in order to support the creation of steering wheels that would elicit particular reactions in either women, men or both.

For example, if the designer wished to design a steering wheel that would engender a benefit such as a feeling of safety, then he or she would input a string such as 'a safe feeling' into the database. This would link to the experiential property associated with a safe feeling (established from analysis of dealer–customer dialogue), which in turn would link to the formal properties associated with this experiential property (established from the cluster analysis). The database would then return images giving design suggestions containing the formal design properties associated with safety.

Kansei Engineering represents a thorough, formal approach to the linkage of product properties with user responses. Like SEQUAM – described in the next section – it relies on statistical analyses to make these links. Such techniques appear to offer the most reliable and valid means of establishing links between a design's formal properties, its experiential properties and the benefits associated with the design. However, when many property dimensions are involved, as in the coffee can example, such methods can become unwieldy and time consuming. In this case seventy-two concepts were designed and respondents were asked to mark over 6,000 scales – rating seventy-two concepts on eighty-six different descriptors each makes a total of 6,192 responses. All this, simply in order to give input to the design of a coffee can, seems excessive!

Another possible criticism of Kansei Engineering is that, in analysing the effect of individual design elements, it is implicitly based on the assumption that a design is the sum of its parts – in other words, the sum of its formal properties. The merit of such an assumption is a matter for debate. It could be, for example, that a 'Gestalt' model is more appropriate. Gestalt theories assert that entities, such as designs, must be considered holistically – that is, to suggest that the overall response to a design could amount to either more or less than the sum of reactions to each of its formal properties.

From the Gestalt point of view more qualitative approaches may be more appropriate, as they may be better suited to looking at the product as a whole. However, it might be argued that, unless the formal properties of the design elements are separated, it is difficult to give designers any meaningful advice as to how to create a product that will deliver particular benefits. Although it may sometimes be unwieldy to apply, Kansei Engineering is arguably the most reliable and valid technique for linking product properties to product benefits. The technique has a proven track record in applications covering a wide range of products and has proved an effective approach to creating designs that will delight the user.

SEQUAM

Sensorial quality assessment (SEQUAM) was developed by Lina Bonapace and Luigi Bandini-Buti at their ergonomics studio in Milan. Like Kansei Engineering, this approach involves analysing a product or prototype in terms of the formal properties of its aesthetic elements and then, through empirical trialling, involving structured interviews, linking these properties with product benefits (Bonapace 1999). SEQUAM differs from Kansei Engineering in that, whilst Kansei Engineering approaches tend to regard formal properties as existing in terms of discrete states, SEQUAM looks at them as variables that exist on a sliding scale. For example, whilst a Kansei analysis might classify a material surface as being, say, rough or smooth, or hard or soft, a SEQUAM analysis might classify a formal property in terms of a particular roughness co-efficient and a particular hardness co-efficient. So, whilst Kansei Engineering relies on cluster analysis for relating product properties to user responses, SEQUAM relies on correlation co-efficients.

To illustrate how SEQUAM works, consider the following example. Imagine that a product development team were working on the design of, for example, a highball glass. Imagine that it had been decided that the product should feel good to hold, that it should feel good against the lips when drinking and that the user should feel elegant when drinking from the glass.

Imagine that the design team considered a highball glass in terms of its formal properties with respect to the following dimensions: weight, height, diameter, glass texture (roughness) and glass thickness. The starting point of the SEQUAM analysis would be to list these pleasure requirements – benefits – and the dimensions of the formal properties that could be manipulated in the design in order to achieve these. The required benefits and the dimensions of the formal properties to be manipulated are summarised in Table 4.1.

The next stage might be to buy a selection of highball glasses and to measure their formal properties with respect to the dimensions to be manipulated. Imagine that the analyst obtained, say, seven highball glasses. He or she would then rank these glasses in terms of their comparative positions on the dimensions to be manipulated.

Having ranked the formal properties of the glasses with respect to each dimension, the analyst might then take measurements of people's subjective responses to each glass. In this case, the aim is to investigate what makes a glass feel good in the hand, what makes it feel good on the lips, what makes it seem elegant and what helps it radiate an air of quality. Here, then, the analyst might design an empirical trial in which a sample of, say, ten people were asked to handle and drink from each of the glasses and then to rate their response to each glass on numerical scales. For each glass, each participant might be asked to mark scales to indicate the extent to which he or she felt that the glass delivered the specified benefit.

From the responses given on these scales, it would then be possible to calculate a mean rating for each glass with respect to the extent to which it was perceived as delivering the specified benefits.

Having gathered data about responses to each of the highball glasses under test and having rated the glasses' formal properties with respect to the dimensions to be manipulated, these can then be correlated against each other in order to investigate associations between the formal properties and the specified benefits.

The correlation co-efficients that emerge give the degree of association between the formal properties of a design and the benefits delivered. The larger the correlation co-efficient, the more strongly the formal property and the associated benefit are linked. How large the co-efficient must be before

Table 4.1 Benefits required of a highball glass and dimensions of formal properties to be manipulated

Benefits	*Dimensions of formal properties*
Feels good in hand	Weight
Feels good on the lips	Height
User feels elegant	Diameter
	Roughness
	Thickness
	Rim curvature

the analyst accepts that there is a meaningful link between a formal property and a benefit is likely to be dependent upon the context in which the analysis is being carried out. If the analysis is within the context of a research study, where the intention is to make definitive, generalisable property–benefit links, then the analyst may decide that only links supported by statistically significant correlations would be accepted as meaningful. On the other hand, if the analysis were to occur in the context of a commercial product creation process, then the analyst is likely to be influencing a design process where decisions – one way or the other – have to be made. In this case, then, the analyst is not necessarily interested in whether a co-efficient is statistically significant, but rather is interested in whether it suggests – on the balance of probabilities and in relation to, for example, cost implications of particular design decisions – that it is worth taking into account.

Originally, SEQUAM was developed in the context of automotive applications in conjunction with Fiat. Bandini-Buti *et al.* (1997) reported the use of the technique in the development of the interior of the Fiat Uno. SEQUAM was used as part of a research programme that ran between 1992 and 1996 in order to investigate the sensorial qualities of car interiors. Amongst the aspects of the interior investigated with the SEQUAM approach were the design of: steering wheels, manual gear levers, automatic gear levers, column-mounted levers, heater controls, ceiling lights, and internal and external door levers and handles.

In the case of the door handle, the analysis was started with the definition of the parameters that were thought to influence the sensorial, tactile, visual and acoustic perception of the door handle – in other words, with a definition of the dimensions of the formal properties that were thought to influence the experiential properties. This was done on the basis of research into the door handles of vehicles currently on the market. After visiting a number of motor fairs and car show rooms, the researchers had a feel for the sorts of design variations that could be important. They then gathered together fourteen cars whose internal door handles varied according to these parameters and asked eighteen people – nine men and nine women – to try these out and comment on them. This was not part of the SEQUAM analysis itself, but rather a preliminary check to investigate which formal properties of door handles were likely to influence user perceptions. In this case, fourteen different parameters were identified. Each of these was related to either the shape of the handle and the handle recess (for example, the length of the lever, the thickness of the lever), the manner in which the lever was deployed (for example, the effort required to deploy the lever, the volume of the sound associated with deployment) or the manner in which the handle was grasped (for example, angle of hand when grasping, wrist rotation during grasping).

Having identified the parameters, the analysts then designed a series of thirteen prototype car door handles, which they mounted on simulated car doors in a specially equipped workshop. The same eighteen participants, who had been involved in identifying which formal properties might be important, were invited back to take part in user trials in which they performed simulated tasks with each of the prototypes and rated them according to various aspects of pleasure (not divulged by the authors) using ten-point numerical scales. This, as in the fictitious example of the highball glasses, enabled correlations to be performed in order to link particular formal properties with particular experiential properties and product benefits.

Bandini-Buti *et al.* end their paper by expressing the intention to create data-banks, so that, over time, it may be possible to establish some general rules and guidelines about the link between formal design properties and product benefits. So far there have been too few studies of this nature done to be able to track any systematic property–benefit links across different product types.

SEQUAM analyses are similar to Kansei Engineering analyses in that they rely on using statistical tests of association in order to link product properties to product benefits. Like Kansei Engineering, SEQUAM is a thorough technique that is robust in that it relies on the analysis of quantitative data gathered via a tightly controlled process. Because SEQUAM sees product properties as lying on continuums and uses correlation as its statistical basis for property–benefit association, comparatively few samples are required for each analysis. Sufficient samples are needed in order to make correlation meaningful – probably a minimum of seven or so. This is as opposed to the clustered-element approach of Kansei Engineering, which requires that formal properties are clustered in various combinations on different samples – potentially requiring the development of vast numbers of prototypes if many different elements are being investigated.

Although this approach may make SEQUAM a more efficient technique than Kansei Engineering in terms of the number of participants required for an analysis, it can make it vulnerable to a number of potential sources of misinterpretation. One potential problem is that it may miss interaction effects. Consider again the example of the highball glass. Imagine that the correlation analysis indicated that narrowness and lightness are correlated with each other and that they are both correlated with the benefit of a feeling of elegance when drinking from the glass. Knowing this, the analyst could not be sure as to whether it would be necessary for the glass to be *both* narrow and light in order to be seen to give a feeling of elegance or whether one or the other of these properties on its own would contribute to the feeling of elegance.

Not knowing this can be a problem when design decisions are made. In the example given, imagine that lightness were detrimental to perceptions about quality and to the pleasantness of the glass in the hand – that is to say, that weight positively correlated to the glass feeling pleasant in the hand. If this were so, the design team might hope to deliver the benefit of elegance simply through the narrowness of the glass. Without knowing about the presence or absence of an interaction effect between lightness and narrowness, it is difficult to predict whether this would be sufficient. An approach based on cluster analysis would have been able to have picked this up as some of the prototypes would have been narrow and heavy and others narrow and light. The participants' responses to these prototypes would then reveal whether both narrowness or lightness were required to make the user feel elegant or whether narrowness alone would be sufficient.

Another potential difficulty is that, whilst it may be appropriate to think of some formal properties as lying on a continuum, this may be less appropriate for others. Properties such as weight, height, thickness, etc., can clearly be seen as points on a continuum; however, for others such as colour and shape this may seem less appropriate. Properties such as colour and shape *could* be seen as lying on a continuum. For example, blue, red and green might be seen as lying at different points on a wavelength continuum. Similarly, a square form and an oval form *could* be seen as shapes that exhibit different degrees of roundness. However, it seems very doubtful as to whether, experientially, people would actually perceive these properties in this manner. This may mean that a correlation-based analysis would not be appropriate for some properties.

Another caution on the interpretation of correlations is to be aware that these are likely to hold within limits, but that the analyst should beware of taking the associations beyond their boundaries. For example, in the case of the highball glasses, narrowness may have emerged as being, in the context of the reactions sought, a desirable property. However, this is likely to be so only within particular limits. Clearly, a glass that was ridiculously narrow would be very difficult to drink from and is unlikely to make the user feel at all elegant! In reality, this issue need not present too much of a problem – it is simply a case of the analyst and the design team using common sense and professional judgement in translating the property–benefit associations into design decisions. In this case, for example, it may be appropriate to create a design that is narrower than most of the competition, but not necessarily the narrowest on the market.

Product Personality Assignment

Product Personality Assignment is an approach developed by Philips Design (Jordan 1997). In this approach 'personality' is seen as an experiential property of a product. The study reported later in this section had a number of aims: to investigate whether the concept of products as personalities was meaningful; to investigate whether there was a link between product personality and people's preferences for products; and to investigate possible links between product personality and the formal properties of a design.

Pleasure-based approaches see products as being 'living objects'. The idea that products could be seen as having their own personalities – that particular personality traits could be seen as part of a product's experiential properties – is an extension of this idea and forms the premise on which Product Personality Assignment is based. This may seem fanciful: after all, products do not have personalities in the strict psychological sense of the word. Nevertheless, people may see products as having personalities – some products seem awkward and unhelpful, some seem calm, and some seem funny and humorous.

Indeed, both anecdotal and empirical evidence suggests that people do project human characteristics on to products. For example, speaking on the BBC's *Top Gear* motoring programme, journalist Jeremy Clarkson described the latest Mercedes as looking 'surprised', 'as if someone had just stuck a banana up its bottom'. Indeed, the common practice of people giving their cars names seems another indication of the attribution of human characteristics to motor vehicles.

An early study on the issue of pleasure with products (Jordan and Servaes 1995) drew on data gathered from case study reports of interviewees' most pleasurable and displeasurable products. Transcripts of the interviews also revealed that people were talking about these products in human terms. One participant, for example, said that he had come to regard his portable radio as an 'old friend'. Another interviewee said that she was sometimes violent towards her stereo – thumping it in order to 'punish' it for damaging cassette tapes.

A follow-up study specifically addressed the issue of how people perceived product personality and which formal properties of products were associated with particular 'personality traits'. The study used an approach, known as Product Personality Assignment, in order to ask a group of participants to assign personality characteristics to a series of electrical consumer goods. The potential personality classifications used in Product Personality Assignment were derived from the work of Briggs-Myers and Myers (1980) that, in turn, builds on Jung's work on human personality types (Jung 1971).

One of the outcomes of the Product Personality Assignment study was to demonstrate that different people tended to assign similar personality characteristics to the same product. That consistent patterns were found in the data appears to suggest that the concept of 'product personality', as an experiential property of a product that could be related to particular formal properties, really is meaningful. Consistent patterns in data appear to signify that people have common perceptions about what a product's personality might be – indicating that product personality is a phenomenon that those involved in product creation should take seriously. This part of the study involved showing participants a selection of products and asking them to indicate, via a questionnaire, which personality traits they felt fitted the product. They did this by selecting from a list of possible traits derived from the Briggs-Myers personality type indicator.

The study also asked about the reasons why people assigned particular personalities to products. This was done in focus group sessions in which each participant explained to the group why he or she had assigned a particular personality to a particular product. During the ensuing conversation the investigator then noted the associations that participants were making between the experiential property of a product's personality and the formal properties of that product. The data indicated, for example, that people tended to rate simple geometrically styled products as being 'sensible' and 'trustworthy'. Meanwhile, products that were designed to be more organic in their style were seen to be more 'friendly' and 'intuitive', even 'cute'. Colours, materials and finishings also had a major influence on the way in which people assigned personalities to products. For example, light colours, metals and smooth, shiny finishes were associated with making a product's personality seem 'extrovert'; dark plastics, on the other hand, were associated with making a product seem 'introvert'.

Another issue, addressed by the study, was the link between a product's personality and the extent to which people reported liking or disliking it. Participants were asked to complete another questionnaire. This time they were asked to give information about which of the products in the study they liked the most. To investigate whether there was any relationship between people's own personalities and their preferences with respect to product personality, participants were also asked to select characteristics that they felt reflected their own personalities.

Overall, the study showed no link of any particular personality characteristic to product preference. In other words, it was not possible to conclude that, for example, 'extrovert' products were preferred to 'introvert' ones. However, participants did show a strong, and statistically significant, preference for products that they felt reflected their own personality. So, for example, if a respondent regarded himself or herself as being 'extrovert' then they were more likely to express a preference for 'extrovert' products. 'Introverts' on the other hand tended to prefer products to which they had assigned 'introvert' personalities. Again, this underlines the importance of understanding people in a holistic manner. An understanding of common personality traits within a particular target group can help those charged with product creation to design a product that is more likely to appeal to their target audience.

Mental Mapping

Sometimes, people may associate a product with a particular person and may project that person's characteristics on to a product – thereby assigning particular experiential properties to it. That is the theory of Stan Gross, a marketing and design consultant to many of America's largest manufacturing corporations. Gross has developed a technique that he calls 'Mental

Mapping' (Hine 1995), which he uses to assess the benefits and drawbacks associated with different design concepts, in particular packaging designs. His research has demonstrated, amongst other things, that people sometimes make links between a design and well-known public figures. For example, interviews and focus groups that Gross has conducted with selected members of the public indicated that the formal properties of the form of the packaging of a particular brand of washing detergent reminded many who saw it of Sylvester Stallone. Accordingly, participants projected properties such as powerfulness to the detergent inside the package. Gross saw this as a positive thing, as it helped reinforce the opinion that the product would 'get the job done' (Hine 1995: 231). On the other hand, Gross was less positive about a toothpaste package design that reminded people of Arnold Schwarzenegger, feeling that people might regard the product as too forceful for its purpose. Here, then, people are making judgements about the experiential properties of a product – 'powerful' in this case – based on an association with a famous person. This association is initially formed due to aspects of the product's formal properties.

Gross contends that about 90 per cent of people's thoughts are not consciously accessible, but rather that they are responses which come directly from what he terms the 'inner mind'. Because of this, he believes that people rarely know what it is that makes them respond positively or negatively to something – therefore it is not appropriate to ask them directly. Gross contends that it is the inner mind which decides on whether a person likes something and that the rational mind simply makes *post hoc* rationalisations of this. If this is so, he argues, then techniques that rely on users responding to rational questions about what they like about a product are inherently doomed to be misleading. This is because they will present the investigator with rationalisations rather than truths about why people respond positively to product features.

There is some support for Gross's contentions in the psychology literature on cognitive dissonance (see Banyard and Hayes 1994 for an overview of this topic). The basic principle behind cognitive dissonance is that people will often try to find a rationalisation for their opinions and preferences, even if these opinions and preferences do not, in fact, have a rational root. This effect is exaggerated when the person has a major emotional attachment to a product. For example, someone who has spent a lot of money on a new car will want to feel justified in that decision. It may well be that he or she bought the car simply because they had a good 'gut feeling' about it. Subconsciously, however, the buyer may want to convince himself or herself that the purchase was rational and thus uses the rational mind to find reasons justifying the purchase. He or she may point out, for example, that the colour scheme is classy, that the excellent acceleration is exhilarating, that the fuel economy is sensible and that the seats are comfortable. Indeed, all of these things may be true. However, they may not necessarily be what attracted the person to the product. Indeed, the person may have no idea what the real attraction was – perhaps it really was these rational elements, but perhaps it was something about the 'personality' of the car, or maybe the design evoked some positive association.

New York Times journalist Thomas Hine describes Gross's methods as 'part Vienna, part Catskills' (Hine 1995: 230). The basis of Mental Mapping is to use popular cultural references to investigate the link between the formal properties of a product and the benefits that people associate with the product. This is based on the premise that the inner mind works on the basis of myths and other sorts of stories, many of which are accessible through popular culture. Gross brings together groups of people – often both designers and potential end users – and sets up games that he claims help illicit responses from the inner mind. They might start out by doing something to help them relax – so far he has made people do anything from playing baseball to drawing with crayons. The aim here is to pull the mind out of its analytical state and to create a sense of playfulness and openness. Indeed, the investigation itself tends to proceed in an equally whimsical way. Gross divides the participants into small groups, gives them products and then asks them to tell stories about the product. For example, he might ask them to imagine that a product was a person who has just died and request that they write its obituary, or he might ask them to imagine that a product was a movie and ask what the plot would be.

Hine (1995) describes the application of Gross's technique to the design of a logo for a new line of food packages. Market research data had shown this logo to be popular and highly recognisable. However, on the basis of his own investigation, Gross contended that the logo was inappropriate to the product line. When he asked people to imagine it as a film plot he found that they were coming up with story-lines that resembled *Fatal Attraction*. He concluded that people were assigning experiential properties to the package design that were too seductive and aggressive for the food contained in the package. (Hine does not report what type of food this was.) On Gross's recommendation, the design was toned down. The product range went on to be a great success.

Gross has also investigated the design of confectionery packaging. For example, when investigating the design of chewing-gum packets he asked people to make up stories about two of the brands currently on the market. Participants were first asked to imagine that the gum was a person and then to give some details of its life. He found that people saw one of the brands as skinny and active and the other as an overweight under-achiever. Details from these 'biographies' are linked to the sorts of responses that people had to the product's packaging. For example, the analogy of the second gum as an 'ill bred person' and its imagined 'poor performance in art and music' (Hine 1995: 232) were seen by Gross as being indicative of the packaging radiating a sense that the product was badly made and that it was not refreshing. Once again, then, the formal properties of a package design were leading people to project negative experiential properties on to the product inside.

Sometimes Gross will ask people to imagine that a product is a person and to make up some stories about its life. If, for example, a product's 'story' included incidents where it cheated others or promised things it did not deliver, then Gross might see this as reflecting a perception of dishonesty in the design – for example, needless over-featuring or deceptively 'flashy' finishings. In one analysis session, participants made up a story about a toothpaste package that seemed to reflect the plot of *The Little Mermaid*. The colour of the package was blue – prima facie perhaps a reason for the underwater associations. However, according to Gross, this was not the main issue. The central point, according to Gross, was that *The Little Mermaid* represented 'someone who is trying to be what she is not…a product that is making too many promises…people won't adopt it as their regular tooth-paste because it seems too much like a novelty' (Hine 1995: 231).

Sometimes, instead of, or in supplement to, the story-telling sessions, Gross will hypnotise people in order to understand what it is that makes them respond to a product in a particular way. Again, this is seen as a way of looking beyond people's 'rational' explanations, to come to an under-standing of the deeper role that they wish the product to play in their lives. In this way, he has often uncovered unconscious motives in a person's rela-tionship to particular products. For example, after hypnotising a group of parents, he found that many felt concern, even guilt, about serving their chil-dren pre-packed frozen meals. He was able to conclude from this that the packaging on this type of food should radiate an air of reassurance with respect to its nutritional value. In other words, the package design should have the experiential property 'reassuring'.

Gross's approach – an 'off-the-wall' application of interviews, focus groups and participative creation – may seem unconventional, perhaps even bizarre. However, it is based on the premise that people's reasons for reacting to products may not be rational, a premise that seems, prima facie, sensible. After all, it seems difficult to analyse many human relationships in purely rational terms or to investigate them effectively using conventional formal methodologies. It might be difficult, for example, for a person to give a rational reason as to why he or she likes, loves or trusts one person and not another – although he or she might try to rationalise this if asked. It seems reasonable to assume that the same difficulties could apply to person–product relationships. Mental Mapping seems to offer a means of diving under rational – or, perhaps, rationalised – explanations to get at the heart of what it is about the formal properties of a product that sparks a particular response.

Despite the unconventionality of his approach, Gross's track record suggests that his investigations do produce useful results. Indeed, he is in huge demand from many of the largest corporations in the USA. Nevertheless, a possible weakness of such an approach is that it relies on a lot of assumptions in the interpretation of participants' comments. For example, it seems, prima facie, that there could be a number of reasons for associating a product with the film, *The Little Mermaid*, in addition to the explanation offered by Gross. It seems, then, that the effectiveness of this method is likely to be somewhat, perhaps highly, dependent on the skill and judgement of the analyst. Perhaps Gross is simply a very insightful man when it comes to judging the effectiveness or otherwise of packaging designs. Perhaps his success has little to do with the approach he employs. Because Gross's approach has been widely applied by other analysts, it is difficult to know whether his success is primarily due to the methodology used, or whether it is simply a reflection of Gross's personal expertise.

The hypnotising of participants raises ethical issues. Many psychologists believe that being hypnotised can be potentially harmful to people and should, as a general rule, only be done when it is an absolutely necessary part of psychological treatment. Certainly, those who hold this view would strongly disapprove of using it on participants in a design process. Second, many would feel that if hypnotism is to be performed, it should only be done by someone who is a trained hypnotist and who knows how to perform it safely, without putting participants at risk. Whilst Gross may be trained in these techniques, the vast majority of human-factors specialists will not be.

Expert case studies

Alastair Macdonald, Head of Product Design at the Glasgow School of Art, has been influential in creating links between the industrial design and human-factors professions through his investigation of the links between products' formal and experiential properties, and the benefits that products bring to those who experience them. In his paper, 'Developing a qualitative sense', Macdonald (1998) reports a number of case studies in which he analyses the link between product benefits and the experiential and formal properties of product design. Macdonald's approach was to look at products that have proved successful on the market and that are commonly regarded as pleasurable to own and use. He then used his judgement as a professional designer in order to analyse the properties that make the products, or particular aspects of products, appealing to people. The methodology, then, is a form of expert appraisal. A selection of his case studies are summarised below.

Karrimor's Condor rucksack

Macdonald cites the properties of this rucksack's buckle as making it particularly pleasurable. The buckle, which has a side release mechanism, closes with a very positive 'click'. The level of resistance required to fasten the buckle also gives it a solid feel, whilst the form, materials and finishings combine to give the buckle an appearance of strength. The formal properties of these elements combine to give a feeling that the buckle is reliable and to generate a feeling of confidence in the user – a psychological pleasure. In this case, then, the formal properties of the buckle – for example, the level of resistance of the buckle clip – are associated with experiential properties such as a solid feel. This, in turn, helps to deliver the benefit of giving the user a feeling of confidence in the product.

Good Grips® kitchen utensils

These products are an excellent example of inclusive design. The handle is moulded from a rubbery thermoplastic called Santoprene®. The sticky properties of this material, along with the ribbing in the handle, makes the handle grippable for those with wet hands. The product is ideal for those who would normally have difficulty gripping, for example those with arthritis. In this case, then, the formal properties of the materials used in the handle contribute to giving the handle the experiential property of being 'grippable'. In turn, this leads to the benefit of the product being easy to use with wet hands and for people who may normally be unable to grip products firmly. Some examples of Good Grips® products are illustrated in Figure 4.6.

Global Knives

These Japanese knives are made from a molybdenum/vanadium stainless steel and are ice tempered in order to give a razor-sharp cutting edge. They have integral handles, which are hollowed out in order to give perfect cutting balance with minimum pressure required. The comfort of the knife in the hand and the tactile sensation of the finely balanced weight contribute to making the knives sensorially pleasurable to hold and to making it possible to cut through foodstuffs with the minimum of effort. Because of their smooth surface and seamless construction, the knives allow no contours for food and germs to collect and thus are exceptionally hygienic. This provides the user with a feeling of reassurance. So, in this case, the formal property 'seamless' contributes to the experiential property 'hygienic', which in turn helps to deliver the benefit of reassurance. Examples of Global Knives products are illustrated in Figure 4.7.

Braun Micron Plus Razor

This is another product that, according to Macdonald (1998), is especially pleasurable because of the materials used in its construction. This electric shaver, designed in 1980 by Dieter Rams, has a body surface of over-

Figure 4.6 Examples of Good Grips® products

moulded soft elastomer thermoplastic polyurethane. This finish, which is moulded on to a polycarbonate body, is decorative, but is also anti-slip and deadens the sound of the shaver's motor. Here, contends Macdonald, is an example of how intelligent and creative use of materials can bring aesthetic, tactile and practical benefits to the user. Indeed, Macdonald notes that recent advances in materials technology have opened up a new set of design possibilities for those involved in the product creation process.

Neen Pain Xenos

This is a drug-free electronic pain-relief device that works through electrical stimulation of the nerve endings. Macdonald asserts that because the proportions of the product's design are a visual reference to the personal stereo, the product can be worn visibly on the body without fear of social stigmatisation. This has been made possible through the use of miniature electronic technology and electronic touch controls – an example of

Figure 4.7   Global Knives kitchen tools

aesthetics and technology combining in order to deliver social benefits in a product that might otherwise have been regarded by some as embarrassing or stigmatising. In this case, then, the experiential property of the form 'personal stereo-like' is associated with the benefit 'non-stigmatising'.

NovoPen$^{(tm)}$

Traditionally, those suffering from diabetes had to use clinical looking syringes and needles. The NovoPen$^{(tm)}$ is a device for the self-administration of precise amounts of insulin. Its appearance is rather like that of a pen – this provides a more positive signal than that of the hypodermic syringe, which is coloured through medical usage and drug abuse associations. This offers the user both ideo- and socio-pleasure, by playing down any stigma that the user and others may associate with syringes and/or the medical condition. The NovoPen[tm] also incorporates tactile and colour codes that refer to the different levels of insulin dosage that may be required. These help to make the product easy to use and contribute to the product's aesthetic profile. The technicalities of administering precise dosages have been translated into easy steps and a discreet but positive click occurs when the dose is prepared for delivery. This provides the reassurance to the user in what might otherwise be a rather daunting task. Finally, the NovoPen[tm] also provides sensorial pleasure through its tactile properties – the pen is shaped to fit the hand comfortably and the surface texture, achieved through spark erosion, is pleasant to the touch. Macdonald cites this product as one in which the technical, aesthetic and functional specifications have been successfully synthesised in order to create a product that, despite its association with an inherently unpleasant task, is, in itself, a pleasure to use. The NovoPen[tm] is illustrated in Figure 4.8.

Samsonite Epsilon Suitcase

Journeys through air terminals can be fraught with stress – both physical and psychological. The three handle options on the Samsonite Epsilon Suitcase provide a number of comfortable options for lifting, tilting or trailing. The handle material is a non-slip rubberised coating that does not become sweaty or slippery in use. These features, help to reduce the stress associated with the situation. The design of the suitcase's wheels allows a controllable and responsive movement in the 'trailing' mode; the suitcase 'obeying' the needs of the user, providing an advantage over other suitcases that do not obey their owners' will.

Figure 4.8   The NovoPen[tm]

Olivetti office machine control p

These control panels for office machines display   ding of the need for reassurance by employing feedback from more than one of the senses. This is an example of good use of the human-factors principle of redundancy (Grandjean 1988). Traditionally, however, the benefit of redundancy – giving user feedback in more than one way – has been cited as being a reduction in user error. It helps to ensure that if users fail to notice or misinterpret feedback given in one way, they will notice it when given in another way. Again, this might be seen as an example of how new human-factors approaches are more holistic. Whilst traditional approaches were focused on eliminating user error, Macdonald has picked up on the emotional need for reassurance when undertaking professional tasks. He notes that the good ergonomics embodied in the keyboard layout are supplemented by the pleasant tactile feedback in the buttons, the appealing visual aspects of the layout and the appealing use of colour. The designers have also paid attention to the audible aspects of the keyboard, which gives a pleasant and reassuring 'clatter' when used.

Mazda MX5 Miata car exhaust

The Mazda team engineered this exhaust so that the sound emitted from it evokes the sound of the classic British sports car. Macdonald cites this as an example of how referential design can evoke reactions associated with another product. In this case he asserts that the link with the classic sports car evokes associations of manliness, youthfulness and success. Interestingly, the Mazda team used Kansei Engineering techniques – discussed earlier in this chapter – in developing this sound. In this case, then, the formal properties of the sound contribute to the experiential property of being 'like a British sports car'. This experiential property brought with it the benefit of helping to make the driver feel masculine, youthful and successful.

Experiential case studies

An alternative case study-based approach was one in which people were asked to describe a pleasurable product and to give an account of the benefits that it brings to them. In another early study of pleasure with product use – reported in Jordan (1999) – a number of case studies were gathered. The intention was to gain an insight into why users associated particular product properties with particular benefits. Six of these case studies are summarised below. Whilst Macdonald (1998) was primarily concerned with linking products' aesthetic qualities to people's responses, this study was more concerned with product properties in general – technical and functional, as well as aesthetic.

These case studies were gathered by interviewing people – in this case mainly students, all residents of Glasgow in Scotland – to describe a product that they owned which they found particularly pleasurable. They were asked about the nature of the pleasure that they got from the product and about what it was about the product that made it so special to them. Other than posing these questions, the interviewer remained passive. The content of what was recorded was, then, steered almost entirely by the interviewees.

Case study 1 – the hairdryer user

This 17-year-old woman chose a hairdryer as her pleasurable product: a product that she described as being 'perfect…the best hairdryer I've ever had'. The reason she was so positive about the hairdryer was because it helped her to style her hair in just the way she wanted. This made her feel attractive and gave her a feeling of self-confidence when she went out. She also mentioned that the hairdryer had an unusual design and was thus something that caught the attention of her friends: 'it's "showy", I like it when people come into my room and see it'.

Here, then, was a link between the product's functional properties – the offer of a number of styling possibilities – and a feeling of user confidence, both in the quality of the product and in the appearance of the user. In particular, the user felt that the inclusion of a three-speed variable motor and a diffuser gave her a far greater feeling of control when styling her hair. The feeling of pride gained when others saw the product was associated with the owner's opinion that the style of the product was noticeably different from others on the market. Indeed, the owner said it looked the same as those used by professional hairdressers – the professional-looking appearance of the product was largely what made it a source of pride to her when others saw it. So, the experiential property 'professional looking' was, for this person, associated with the benefit of feeling proud.

Interestingly, this was the second of this model of hairdryer that she had owned, the previous one having broken down after only two years of use. Nevertheless, she had had no hesitation in buying an identical model the next time around. This, then, is an example of a product whose functional and aesthetics qualities outweighed – in the mind of this user at least – its technical shortcomings.

Case study 2 – the guitar player

The guitar player was a 26-year-old man who played the electric guitar. Again, the guitar facilitated social pleasure. He regarded it as a 'status symbol, particularly amongst people who know about these things'. It had also provided a talking point as it had belonged to Lloyd Cole – lead singer of Lloyd Cole and the Commotions, a Glaswegian band who had a number of hits in the 1980s. He also found playing the guitar an exciting activity in itself. Even just having the guitar near him gave him a feeling of reassurance.

In this case, the technical reliability of the product had enabled the user to form a bond with the product that accentuated the pleasure gained from the functional benefits accrued from the quality of the sound of the guitar. The combination of the functional and technical quality of the product, plus the knowledge that it had been used by acknowledged and successful musicians, had contributed to its positive influence on the person's self-esteem.

The guitarist also praised the aesthetic qualities of the product, saying that it had a very solid look and feel. The use of maple wood in the neck of the guitar was both visually and tactilely pleasing to him. The heavy weight of the guitar and the smooth, solid form contributed to a feeling of confidence in the product and reinforced his view of the product's reliability.

Case study 3 – the video cassette recorder (VCR) user

She was a 26-year-old living alone who rented a VCR that she described as being a 'standard video'. She described the emotional benefits that she gained from the video as being a feeling of anticipation – looking forward to watching what she had recorded – and freedom – not having to stay home in order to catch her favourite programmes.

This case study demonstrated how the functional benefits of the product had proved particularly pleasurable within the context of the user's lifestyle. For this person, the product was a 'service carrier' – something that brought pleasure into her life through the benefits of its functionality. There was no indication of an emotional bond to the product itself, as there had been in the previous two case studies.

Case study 4 – the stereo owner

This 19-year-old man had bought the stereo two years ago. He felt that the stereo offered a reasonable level of functionality that afforded him most of the options which he wanted. However, what made the product particularly pleasurable was a combination of aesthetic, technical, functional and usability elements.

He particularly liked the colour of the stereo – black – and felt that the rounded form of the product gave it a certain sophistication. The enthusiasm for black in stereos was mirrored in two other case studies (not reported here). Based on the responses in these case studies, it seemed that black was associated with the experiential properties 'seriousness' and 'sophistication'. These experiential properties can be reassuring to people in the context of products that cost a lot of money to buy. Most of the people whose experiences were investigated in this set of case studies were students.

For many of them, their stereo was the most expensive product that they owned. They were, then, looking for designs that were reassuring and serious. The use of black was associated with this – indeed one man explicitly said that he would not buy a stereo unless it was black, as other colourings made the product seem frivolous.

Case study 5 – another hairdryer owner

This 19-year-old woman had been given the hairdryer as a present eleven years ago. She found the product particularly pleasurable for two reasons. Like the first hairdryer owner, she said that she could use the dryer to style her hair in the way that she wanted and thus it gave her a feeling of confidence in her appearance. Unfortunately, from the data that the analyst recorded, it was not clear which property of the dryer had particularly contributed to making it so suitable for styling; presumably it was some combination of technical performance, functionality and usability, but this wasn't expressed clearly.

In addition, though, she also noted that the product's reliability gave her a feeling of confidence in the product itself. Here, then, there is a link between the experiential property of product reliability and the emotional benefit of a feeling of confidence in the product. That the product had proved technically reliable over a period of many years had enabled the owner to establish a 'bond' with the product – because it had not let her down, the pleasures that she gained from the product were accentuated over an extensive period of use.

This hairdryer owner also noted that the product was small and compact. She felt that this made the product more convenient as she was easily able to pack the dryer into her overnight bag – ideal if she was visiting friends. This, then, is a simple example of an experiential property 'compactness' – which probably resulted from a combination of technology and the formal properties of the aesthetics – being linked to a context of use benefit.

Case study 6 – the television watcher

This 19-year-old woman chose her parents' television as her pleasurable product. She described this as being big and straightforward with an easy-to-use remote control. She said that she particularly liked the television because of its simplicity, because of the large screen and the good picture quality, and because of the wide variety of programmes that she could choose from – her parents had a cable television subscription. Again, though, she did not link these qualities to any particular benefit, other than indicating that, taken together, they led to her feeling 'satisfied' with the television.

In addition, however, she mentioned that the television was very reliable and that this enabled her to take the television for granted. Here, then, was a link between the experiential property 'reliability' and the benefit of a feeling of 'security' that nothing would go wrong.

She also mentioned that the television was black and that this gave an impression of higher quality and reliability – this, then, was an example of how positive experiential properties of the product were reinforced by one of the product's formal aesthetic properties. She also felt that because the product was black it blended in well with its surroundings – her parents' living room. This is as opposed to brighter colours, which may have stood out more.

An advantage of experiential case studies is that they give a direct insight into why users find particular products pleasurable. Although some interpretation is required in order to make the links between the pleasures that users describe and the properties of the products that they mention, this approach relies on far fewer assumptions than does the expert case study approach. Because it is based on direct reports from participants, this approach is useful in identifying the issues that are really important to people; this is as opposed to expert case studies, which rely on the investigator's assumptions about the nature of pleasures associated with particular products. Similarly, participants mention the properties of a product that they find important, whereas with expert case studies it is, once again, the investigator who must make assumptions about these.

On the other hand, there are also a number of disadvantages associated with eliciting unstructured reports about what people find pleasurable in products. One of these is that the method relies on participants' unprompted articulation of the issues that are important to them and the product characteristics associated with these. There are two difficulties here. The first is that participants may not be able to describe exactly what it is that they like about a product – indeed, they may lack the technical and aesthetic vocabulary to do this. This might lead to imprecise descriptions that fail to capture the real essence of a product's appeal. One stereo owner, for example, described his stereo as being 'rounded' – an aspect that particularly appealed to him. To a designer, however, this is a very ambiguous statement. 'Rounded' could mean anything from a radius on the corners, to organic form language, to cylindrical, to spherical. A comparative advantage of the expert case study method is that the analyst has the specialist knowledge to be able to express such issues with precision.

Another disadvantage is that using experiential case studies as a basis for a design may lead to 'lowest common denominator' design. Much of what is reported in the case studies above seems rather predictable – black is a good colour for stereos and television, video cassette recorders should be able to record reliably, hairdryers should be portable. The studies reported above may not necessarily be a rich source of new design ideas; indeed, they may seem simply to confirm what any reasonably thoughtful designer might have predicted about what would make particular products pleasurable to use.

Despite these criticisms, experiential case studies retain the advantage of being a direct, unprompted expression of the pleasures that people can gain from products and the properties of a product associated with these. Before those involved in the product-creation process get carried away with their own – perhaps sometimes fanciful – ideas about what will make products pleasurable, it may be sensible to listen to users in this way. Whilst it would probably be limiting to use experiential case studies as the only basis for linking pleasure to product properties, they may form a very useful supplement to other approaches. They are a good guide to users' basic concerns. Good designs may build imaginatively on these concerns, rather than showing imagination *at the expense of* these concerns.

Sensual pager

Hofmeester and his associates (Hofmeester *et al.* 1996) report a study in which they created a design concept for a pager. The aim was to create a design that would be particularly sensually pleasing, both in the hand and, when carried, against the body of the pager user. Hofmeester felt that the benefit of an experience of sensuality was important in the context of pagers. He asserted that, because sensual experiences are experienced through the body, benefits associated with sensuality are particularly salient to pagers as they are often worn close to the body.

Hofmeester began his investigations by inviting a group of people into the design studio to talk about products that they found particularly sensual. They were asked to bring in these products, or, if the product could not be brought in – for example if it was too big – to bring along a photograph. They were then asked to talk about these products, explaining what it was that made them so sensual and to describe the sensual feelings associated with these products. This was done in the format of a focus group, with participants sharing and discussing their experiences with each other.

Hofmeester then initiated a participative creation session – giving the participants clay and asking them to mould it into forms that they regarded as being sensual. Participants were then asked to describe the sensual objects that they had made in terms of their formal properties. Furthermore, they were asked to describe the nature of the sensuality that the objects facilitated.

Hofmeester made an inventory of the design properties and sensorial benefits mentioned during the interviews and participative creation sessions. He used the sensorial benefits mentioned as the basis for designing a questionnaire for the measurement of sensuality. Meanwhile, in a second focus group he discussed the inventory of the mentioned product properties with the participants and asked them to rate the importance of each property with respect to its overall contribution to a product's sensuality.

A statistical technique – cluster analysis – was used to reduce participants' verbalisations about the design elements affecting perceptions of sensuality into six properties of design. These were as follows. Hofmeester's analysis did not make a distinction between formal and experiential product properties – properties of both types are included in the list below.

Organic form

The analysis suggested that the basic form of the product should be organic, mirroring form elements present in the human body. This meant, for example, that edges should be rounded or curved rather than angular or geometric in form.

Warm

The material from which the product was to be made should feel warm to the touch. This might mean, for example, that materials such as plastics, rubbers or laminates would be more appropriate than colder materials such as metals, ceramics or glass.

Smooth

The material surface should have a low roughness co-efficient. This would suggest that gloss surface finishes would be preferable to matt finishes or knurled surfaces.

Soft

The material of which the pager was to be made should have some 'give' in its surface. Again, rubbers and silicon laminates might be suitable in this respect, as opposed to harder materials such as plastics.

Skin contact

The design of the pager should support the product being worn directly against the skin. Thus it should be wearable under clothing.

Wear against the chest

Responses suggested that a pager which was designed to be worn in contact with the chest would give a greater feeling of sensuality than a pager which was designed for wear on, for example, the arm or wrist.

Hofmeester then designed two pagers, each incorporating a number of the above properties, and conducted a user evaluation in which these were compared against a reference pager. This reference pager was one that Hofmeester and his associates felt was fairly typical of the sorts of pagers currently available on the market. In the evaluation, participants were given the opportunity to hold and wear models of each of the specially designed pagers as well as a model of the reference pager. They were then asked to complete a number of semantic differential scales indicating their response to each of the pagers. The terminology used on these scales reflected the verbalisations in the earlier sessions relating to participants' sensual experiences with products.

An analysis of the results showed statistically significant higher ratings for the new designs on a number of the differential scales.

5

CONCLUSIONS

In this introductory text, an overview has been given of a new approach to human factors. This approach is about fitting products to people in a holistic manner. It is based on a recognition that the quality of the relationship between people and products depends on more than simply product usability. People are more than just 'users'. They have hopes, fears, dreams, aspirations, tastes and personality. Their choice of products and the pleasure or displeasure that products bring to them may be influenced by these factors.

The first chapter started with a review of the recent history of human factors, charting the rise of usability as a hot commercial issue. It was argued that usability has subsequently become something that customers take for granted and has, in marketing terms, changed from being a 'satisfier' to a 'dissatisfier'. This suggests that, if human-factors approaches are to add value to the design process, they must move beyond usability to address the aspects that make products a positive joy to experience. Indeed, it was argued that usability-based approaches to design are – in effect, if not in intention – dehumanising. This is because such approaches tend to encourage a view of people as simply cognitive and physical processors in a user–product–task system. Pleasure-based approaches, on the other hand, encourage a holistic view of users, striving to gain a rich understanding of human diversity.

In the second chapter, a framework – the four pleasures – was outlined. This provides a structured way of approaching the issue of pleasure with products. The framework was illustrated with a series of examples, showing how different types of product can provide people with different types of pleasure. It was argued that, whilst some of these products were pleasurable through coincidences of product history or context, the pleasure associated with many products could be linked to aspects of the products' design. In the third chapter, a structured approach to creating pleasurable products was outlined. There were four main stages to this approach. They were as follows: understanding the people for whom the product is to be designed; understanding the practical, emotional and hedonic benefits that these people would want to gain from a particular product; linking these benefits to the properties of the product design; and evaluating design solutions to check that they can deliver the required benefits. In the fourth chapter, a series of methods, each of which could be used in the product-design process at one or more of these four stages, was discussed. The chapter also reviewed a series of case studies from the human-factors literature in which some of these methods had been applied in the context of pleasure-based approaches to design.

Consumers are becoming more and more sophisticated and demanding. Whilst once people may have been delighted with a product simply because it looked okay and was easy to use, those days are now over. People are looking for well-thought-out designs that show an understanding of their values and lifestyles and will be disappointed with products that fall short in this way. Those who manufacture such products will also be disappointed, as they will soon find that their customers will start to look elsewhere. Through developing holistic approaches to people and design, human factors have much to contribute to the creation of a new generation of well-designed products for people to enjoy.

Pleasure-based approaches are about *really* understanding people and respecting and celebrating human diversity. They are about understanding the benefits that people want from a product and understanding what is required in order to deliver these benefits. Above all, pleasure-based approaches are about designing products that can bring a real joy into people's lives.

REFERENCES

Bandini-Buti, L., Bonapace, L. and Tarzia, A. (1997) 'Sensorial quality assessment: A method to incorporate perceived user sensations in product design. Applications in the field of automobiles', in *IEA '97 Proceedings*, Helsinki: Finnish Institute of Occupational Health, pp. 186–9.

Banyard, P. and Hayes, N. (1994) *Psychology: Theory and Application*, London: Chapman & Hall.

Barnard, P. and Marcel, T. (1984) 'Representation and understanding in the use of symbols and pictograms', in R.S. Easterby and H.J.G. Zwaga (eds) *Information Design 1984*, Colchester: John Wiley & Sons, pp. 37–75.

Beagley, N.I. (1996) 'Field based prototyping', in P.W. Jordan, B. Thomas, B.A. Weerdmeester and I.L. McClelland (eds) *Usability Evaluation in Industry*, London: Taylor & Francis, pp. 95–104.

Bonapace, L. (1999) 'The ergonomics of pleasure', in W.S. Green and P.W. Jordan (eds) *Human Factors in Product Design: Current Practice and Future Trends*, London: Taylor & Francis, pp. 234–48.

Bourdieu, P. (1979) *Distinction: A Social Critique of the Judgement of Taste*, London: Routledge & Kegan Paul.

Briggs-Myers, I. and Myers, P. (1980) *Gifts Differing*, California: Consulting Psychologists Press.

Brigham, F. (1998) 'International standardisation of graphical symbols for computer products', in M.A. Hanson (ed.) *Contemporary Ergonomics 1998*, London: Taylor & Francis, pp. 8–12.

Brooke, J. (1996) 'SUS – A quick and dirty usability scale', in P.W. Jordan, B. Thomas, B.A. Weerdmeester and I.L. McClelland (eds) *Usability Evaluation in Industry*, London: Taylor & Francis, pp 189–94.

Bryson, B. (1995) *Notes from a Small Island*, London: Black Swan Books.

Clarke, S., Jordan, P.W. and Cockton, G. (1995) 'Applying Aristotle's theory of poetics to design', in S.A. Robertson (ed.) *Contemporary Ergonomics 1995*, London: Taylor & Francis, pp. 139–44.

Coleman, R. (1999) 'Inclusive design – design for all', in W.S. Green and P.W. Jordan (eds) *Human Factors in Product Design: Current Practice and Future Trends*, London: Taylor & Francis, pp. 159–70.

Crozier, R. (1994) *Manufactured Pleasures*, Manchester: Manchester University Press.

Dormer, P. (1993) *Design Since 1945*, London: Thames & Hudson.

The Economist (1998) *The Pocket World in Figures*, London: *The Economist*.

Edgerton, E.A. (1996) 'Feature checklists: A cost effective method for "in the field" usability evaluation', in P.W. Jordan, B. Thomas, B.A. Weerdmeester and I.L. McClelland (eds) *Usability Evaluation in Industry*, London: Taylor & Francis, pp. 131–7.

REFERENCES

Edgerton, E.A. and Draper, S.W. (1993) 'A comparison of the feature checklist and the open response questionnaire in HCI evaluation', *Computing Science Research Report GIST-1993–1*, University of Glasgow: Department of Computing Science.

Forty, A. (1986) *Objects of Desire. Design and Society 1750–1980*, London: Thames & Hudson.

Freudenthal, A. (1997) 'Testing new design guidelines for all ages, especially menu-design on home-equipment', in *IEA '97 Proceedings*, Helsinki: Finnish Institute of Occupational Health, pp. 138–40.

Fulton, J. (1993) 'Physiology and design: New human factors', *American Center for Design Journal* 7 (1): 7–15.

Grandjean, E. (1988) *Fitting the Task to the Man*, London: Taylor & Francis.

Green, B. and Jordan, P.W. (1999) 'The future of ergonomics', in M.A. Hanson, E.J. Lovesey and S.A. Robertson (eds) *Contemporary Ergonomics 1999*, London: Taylor & Francis, pp.110–14.

Greenberg, C. (1995) *Mid-Century Modern*, New York: Harmony Books.

Hardy, T. (1998) 'Zen and the art of motorcycle branding', *Innovation*, Summer 1998, p. 55.

Hart, S.G. and Staveland, L.E. (1988) 'Development of the NASA-TLX (Task Load Index): Results of empirical and theoretical research', in P.A. Hancock and N. Meshkati (eds) *Human Mental Workload*, North Holland: Elsevier, pp 139–83.

Hartevelt, M.A. and Vianen, E.P.G. van (1994) 'User interfaces for different cultures: A case study', in *Proceedings of Human Factors and Ergonomics Society Conference 1994*, California: Human Factors and Ergonomics Society, pp. 370–3.

Hauffe, T. (1998) *Design: A Concise History*, London: Laurence King.

Hine, T. (1995) *The Total Package*, Boston: Back Bay.

Hjelle, L.A. and Ziegler, D.J. (1981) *Personality Theories*, London: McGraw-Hill.

Hofmeester, G.H., Kemp, J.A.M. and Blankendaal, A.C.M. (1996) 'Sensuality in product design: A structured approach', in *Proccedings of CHI '96*, New York: ACM, pp. 428–35.

Hofstede, G. (1994) *Cultures and Organisations*, London: HarperCollins.

Ishihara, S., Ishihara, K., Tsuchiya, T., Nagamachi, M. and Matsubara, Y. (1997) 'Neural networks approach to Kansei analysis on canned coffee design', in *IEA '97 Proceedings*, Helsinki: Finnish Institute of Occupational Health, pp. 211–13.

ISO DIS 9241–11. Ergonomic requirements for office work with visual display terminals (VDTs): Part 11: Guidance on Usability.

Johnson, G.I. (1996) 'The usability checklist approach revisited', in P.W. Jordan, B. Thomas, B.A. Weerdmeester and I.L. McClelland (eds) *Usability Evaluation in Industry*, London: Taylor & Francis, pp. 179–88.

Jordan, P.W. (1999) 'Pleasure with products: Human factors for body, mind and soul', in W.S. Green and P.W. Jordan (eds) *Human Factors in Product Design: Current Practice and Future Trends*, London: Taylor & Francis, pp. 206–17.

—— (1998) *An Introduction to Usability*, London: Taylor & Francis.

—— (1997) 'Products as personalities', in M.A. Hanson (ed.) *Contemporary Ergonomics 1997*, London: Taylor & Francis, pp. 73–8.

—— (1994) 'Focus groups in usability evaluation and requirements capture: A case study', in S. Robertson (ed.) *Contemporary Ergonomics 1994*, London: Taylor & Francis, pp. 449–53.

—— (1993) 'Methods for user interface performance measurement', in E.J. Lovesey (ed.) *Contemporary Ergonomics 1993*, London: Taylor & Francis, pp. 451–60.

Jordan, P.W. and Engelen, H. (1998) 'Sound design for consumer products', in *Proceedings of Stockholm, Hey Listen!*, Stockholm: Royal Swedish Music Academy, pp. 73–9.

Jordan, P.W. and Johnson, G.I. (1993) 'Exploring mental workload via TLX: The case of operating a car stereo whilst driving', in A. Gale (ed.) *Vision in Vehicles IV*, North Holland Elsevier, pp. 255–62.

Jordan, P.W. and Kerr, K.C. (1999) 'Pleasure, usability and telephones', in J.M. Noyes and M. Cook (eds) *Interface Technology: The Leading Edge*, Baldock, UK: Research Studies Press, pp. 229–43.

Jordan, P.W. and Macdonald, A.S. (1998) 'Pleasure and product semantics', in M.A. Hanson (ed.) *Contemporary Ergonomics 1998*, London: Taylor & Francis, pp. 264–8.

Jordan, P.W. and Servaes, M. (1995) 'Pleasure in product use: Beyond usability', in S. Robertson (ed.) *Contemporary Ergonomics 1995*, London: Taylor & Francis, pp. 341–6.

Jung, C.G. (1971) *Psychological Types*, trans. H.G. Baynes, Princeton: Princeton University Press.

Kemp, J.A.M. and Gelderen, T. van (1996) 'Co-discovery exploring: An informal method for iteratively designing consumer products', in P.W. Jordan, B. Thomas, B.A. Weerdmeester and I.L. McClelland (eds) *Usability Evaluation in Industry*, London: Taylor & Francis, pp. 139–46.

Kerr, K.C. and Jordan, P.W. (1994) 'Evaluating functional grouping in a multi-functional telephone using think-aloud protocols', in S.A. Robertson (ed.) *Contemporary Ergonomics 1994*, London: Taylor & Francis, pp. 437–42.

Kim, J. and Moon, J.Y. (1998) 'Designing towards emotional usability in customer interfaces – trustworthiness of cyber-banking system interfaces', *Interacting With Computers* 10: 1–29.

Kirakowski, J. (1996) 'The software usability measurement inventory: Background and usage', in P.W. Jordan, B. Thomas, B.A. Weerdmeester and I.L. McClelland (eds) *Usability Evaluation in Industry*, London: Taylor & Francis, pp. 169–77.

Kirakowski, J. and Corbett, M. (1988) 'Measuring user satisfaction', in D.M. Jones and R. Winder (eds) *People and Computers IV*, Cambridge: Cambridge University Press, pp. 329–38.

Kirkham, P. (1996) *The Gendered Object*, Manchester: Manchester University Press.

Kline, P. (1993) *Personality: The Psychometric View*, London: Routledge.

Laurel, B. (1991) *Computers as Theatre*, London: Addison-Wesley.

Lewis, C.S. (1960) *The Four Loves*, London: Fount.

McCormick, E.J. and Sanders, M.S. (1983) *Human Factors in Engineering and Design*, fifth edn, Auckland: McGraw-Hill.

Macdonald, A.S. (1999) 'Aesthetic intelligence: A cultural tool', in M.A. Hanson, E.J. Lovesey and S.A. Robertson (eds) *Contemporary Ergonomics 1999*, London: Taylor & Francis, pp. 95–9.

—— (1998) 'Developing a qualitative sense', in N. Stanton (ed.) *Human Factors in Consumer Product Design and Evaluation*, London: Taylor & Francis, pp. 175–91.

Macdonald, A.S. and Jordan, P.W. (1998) 'Human factors and design: Bridging the communication gap', in M.A. Hanson (ed.) *Contemporary Ergonomics 1998*, London: Taylor & Francis, pp. 551–5.

McKellar, S. (1996) 'Guns: The last frontier on the road to equality?', in P. Kirkham (ed.) *The Gendered Object*, Manchester: Manchester University Press, pp. 70–9.

Marzano, S. (1998) *Thoughts*, Blaricum, the Netherlands: V+K Publishing.

—— (1993) 'Flying over Las Vegas', *EDC News*, September 1993: 1–33.

Maslow, A. (1970) *Motivation and Personality*, second edn, New York: Harper & Row.

Maulsby, D., Greenberg, S. and Mander, R. (1993) 'Prototyping an intelligent agent through Wizard of Oz', in *Proceedings of INTERCHI '93*, New York: ACM, pp. 272–84.

Mooij, M. de (1998) 'Mapping cultural values for global marketing and advertising', *International Journal of Marketing*: 681–703.

Nagamachi, M. (1997) 'Requirement identification of consumers' needs in product design', in *IEA '97 Proceedings*, Helsinki: Finnish Institute of Occupational Health, pp. 231–233.

—— (1995) *The Story of Kansei Engineering*, Tokyo: Kaibundo Publishing.

Norman, D.A. (1988) *The Psychology of Everyday Things*, New York: Basic Books.

Noyes, J.M. (1983) 'The QWERTY keyboard: A review', *International Journal of Man–Machine Studies* 18: 265–81.

O'Donnell, P.J., Scobie, G. and Baxter, I. (1991) 'The use of focus groups as an evaluation technique in HCI', in D. Diaper and N. Hammond (eds) *People and Computers VI*, Cambridge: Cambridge University Press, pp. 211–24.

Paglia, C. (1995) *Sex and Violence, or Nature and Art*, London: Penguin Books.

Pheasant, S. (1986) *Bodyspace: Anthropometry, Ergonomics and Design*, London: Taylor & Francis.

Popcorn, F. (1996) *Clicking*, London: Thorsons.

Ravden, S.J. and Johnson, G.I. (1989) *Evaluating Usability of Human–Computer Interfaces: A Practical Method*, Chichester: Ellis Horwood.

Scott, G. (1995) *Motorcycle Mania*, London: Carlton.

Shackel, B. (1999) 'How I broke the Mackworth Clock Test (and what I learned)', in M.A. Hanson, E.J. Lovesey and S.A. Robertson (eds) *Contemporary Ergonomics 1999*, London: Taylor & Francis, pp. 193–7.

Shiotani, Y. (1997) 'Canon Elph: The power of collaboration', *Innovation*, winter 1997: 94–7.

Soloman, M.R. (1996) *Consumer Behavior*, third edn, New Jersey: Prentice-Hall.

Steele, V. (1997) *Fifty Years of Fashion*, New Haven: Yale University Press.

Strinati, D. (1995) *An Introduction to Theories of Popular Culture*, London: Routledge.

Tambini, M. (1996) *The Look of the Century*, London: Dorling Kindersley.

Tiger, L. (1992) *The Pursuit of Pleasure*, Boston: Little, Brown & Company, pp. 52–60.

Vermeeren, A.P.O.S. (1996) 'Getting the most out of "quick and dirty" usability evaluation', in P.W. Jordan, B. Thomas, B.A. Weerdmeester and I.L. McClelland (eds) *Usability Evaluation in Industry*, London: Taylor & Francis, pp. 121–8.

Virzi, R.A. (1992) 'Refining the test phase of usability evaluation: How many subjects is enough?', *Human Factors* 34: 457–68.

Votolato, G. (1998) *American Design in the Twentieth Century*, Manchester: Manchester University Press.

Vries, G. de, Hartevelt, M. and Oosterholt, R. (1996) 'Private Camera Conversation method', in P.W. Jordan, B. Thomas, B.A. Weerdmeester and I.L. McClelland (eds) *Usability Evaluation in Industry*, London: Taylor & Francis, pp. 147–55.

Vries, G. de and Johnson, G.I. (1999) 'Spoken help for a car stereo: An exploratory study', in W.S. Green, and P.W. Jordan (eds) *Human Factors in Product Design: Current Practice and Future Trends*, London: Taylor & Francis, pp. 124–37.

Woodham, J.M. (1997) *Twentieth-Century Design*, Oxford: Oxford University Press.

Zajicek, M., Powell, C. and Reeves, C. (1999) 'Ergonomic factors for a speaking computer interface', in M.A. Hanson, E.J. Lovesey and S.A. Robertson (eds) *Contemporary Ergonomics 1999*, London: Taylor & Francis, pp. 484–8.

INDEX